HACKING THEOLOGY

HOW TECHNOLOGY REVEALS GOD TO US

HACKING THEOLOGY
HOW TECHNOLOGY REVEALS GOD TO US
Marcus Guevara

Thirsting for Truth
www.ThirstingforTruth.com

First paperback edition December 2018

ISBN: 978-0-578-43721-7 (paperback)
ISBN: 978-1-7335392-0-3 (ebook)

Published by Thirsting for Truth
www.ThirstingforTruth.com

No one ever gets anywhere in life without the help of others. I've had so many people contribute to my life and my successes. I'd first like to thank my parents, Clovis and Cecilia, who were the first to love me in this world. You recognized my love for technology early on and you supported me through every step of my life and career. I remember how much you used to brag to others about me and my knack for computers. You used to tell people that I could do all these amazing things that I actually didn't know how to do. Your constant confidence made me want to strive to deserve the praise you gave me. You purchased so many books to support my habit for reading. You bought me computers to support my desires for technology. You encouraged me, pushed me, and challenged me to be better as a person. This book is largely a product of your investment in me. Thank you for spending the time to help mold me into who I have become. To my wife, Dianne, and my kids, Elizabeth, John-Paul, Anna Marie, Gabriel, and the one-not-yet-named: thank you for sacrificing time with me so often to let me finish this book. Thank you for the way you love me. You give me inspiration to do great things. I'm so blessed to have such an awesome family.

Σ Η Μ Ι Λ Υ Α

Contents

WHY I HACKED THEOLOGY
(AND WHY YOU SHOULD TOO)

It's 11:30 pm and I'm sitting in front of the computer screen in the living room. I'm going back and forth between Half-Life (a video game), mIRC (a chat program) and AOL Instant Messenger. I'm having a pretty awesome night until my mom walks into the living room and shouts my name. "Marcus! What are you doing?!"

"How did she know I was here?" I thought.

The thing is, I was only in the sixth grade at the time. I had waited for my parents to go to sleep before I came out of my room to begin my nightly activities. This wasn't my first school night staying up late on the computer while my parents believed I was sleeping, and it wouldn't be my last. This wasn't just video game time for me, though. This was where I learned some valuable stuff that paved the path toward a very successful career. By seventh grade, I had taught myself programming in Visual Basic and C++. I became part of online groups where I learned cool tricks, and eventually found hacking tools I could use to impress friends. By eighth grade, I was selling burned copies of music, movies, and video games to my fellow classmates. This was in 1999 or 2000 and pirating was not yet a huge issue because the only way to get a hold of the kind of content I had was through newsgroups and internet relay channels. You had to be invited and be able to use a command-line interface (a programming language of sorts) so most of the population was ignorant of these large libraries of entertainment. Still, while the crusade against pirating was not yet in full force, dads were still dads and mine happened to find a wad of cash in my backpack one day which busted my whole operation. My dad still loves to tell that story around holidays and basically anytime the family is together. As a disclaimer, I don't condone pirating (or any illegal activity). I don't tell that story to encourage pirating but to highlight the fact that I wasn't your average thirteen-year-old kid. While I played sports all my life and liked to dance and play the guitar, my real passion was computers and video games. I could sit in front of a screen for hours on end playing, learning, and creating. I was so fascinated that I would often get lost in the digital world. This became a point of tension with my parents, who wanted to support my passion for computers but were concerned with how much time I spent in front of the monitor. Naturally, I believed any sanctions they implemented were unfair and I began to find ways to circumvent them. Thus, I would try and sneak into the living room when everyone was asleep to get in some extra time or indulge after school before my parents came home from work. As you can imagine, that caused many fights. At times my mom would get so mad that she would disconnect the computer and put it in her closet (I would reconnect it and use it when she wasn't around).

The great thing about my parents, though, is that they supported my passion far more than they implemented sanctions out of their concern. I still remember one Christmas in middle school when they bought me my first computer books, "Java for Dummies" and "Visual Basic 6 for Dummies". It was 1998 and I was eleven years old. My parents didn't have a lot of money and those books cost them $50. That's a lot of money for a Christmas gift in 2017; so, in 1998, it was significant. Two years later, my parents bought me a computer that had eight gigabytes of hard drive space and cost a couple thousand dollars. This is a testament to the love my parents had for me. They were always my greatest supporters even though they didn't understand technology and knew there were potential dangers. They recognized my passion and did what they could to foster it. They knew—just like I did—that computers would be a big part of my life forever. Never during high school did I struggle with confusion over what I would study in college and what I would do for a career. I knew that I would simply continue to do what I loved to do and so I did. In 2009, I graduated with a bachelor's degree in Computer Science. I immediately landed a job as a computer programmer and I loved it. I decided to continue my education and went on to obtain a master's degree in Cybersecurity. After that I joined the Air Force as a reservist and then spent a few years working as a contractor performing vulnerability assessments and penetration testing. I would go on to become an officer in the US Coast Guard, continuing to help defend the nation from cyber-attacks. I am still working in the cyber field today and I owe a great deal of my success to my parents and their support. I still have the books my parents gave me in 1998. I keep them as a reminder that I didn't get here on my own. I hope that, when the time comes, I can support my kids' dreams the way they supported mine.

Sweet story, right?

Well, unfortunately that's not the whole story. The truth is that while computers have brought great joy to my life, they have also been the source of much pain and turmoil. I certainly learned a lot as a kid and did some very cool, and creative, things but I also got involved in a lot of bad things. I spent a lot of time getting to know people in underground groups and chat rooms where I was exposed to some pretty horrific stuff. People sharing free music and movies were also sharing and promoting pornography and other extreme forms of "entertainment". At twelve or thirteen years old, I remember being exposed to twisted forms of hardcore pornography and disturbing pictures, and videos, of accidental deaths and murders. There were websites and services dedicated to hosting stolen crime scene photos and horrific crimes captured on video. There are many other terrible examples of the kind of evil that lurks in these circles, but the point is that my time on the computer was often destroying my heart and mind as much as, or even more than, it was building them up. I also developed an unhealthy addiction to computers and video games which brought about its own set of problems. Thankfully, my parents didn't always make it easy for me to misuse the computer. They supported me, but they also tried to limit me, talk to me, and they punished me when I was not following the rules. They also kept me involved in sports, outdoor activities, and forced me to occasionally find things to do at home that didn't involve the computer. Without the constant tension, and supervision, I received from my parents I might have fallen much deeper into bad habits and my life could be very different right now. There's a lot more to my story but the point is that one could argue that technology has been a great good in my life just as easily as they could argue that it has caused destruction. This is a big reason why many parents struggle with the question of whether they should support their kids' desires for technology or be concerned about them. While technology has changed drastically in the past 20 years, our constant state of confusion regarding whether or not technology is ultimately good has not changed. I believe I have the answer to this question and, the funny thing is, I actually found it by accident.

I didn't start practicing the Christian faith until I was in college. My conversion story— which is something I now travel the nation sharing— is powerfully expressed in a popular YouTube video titled, "Worth the Wait". You can find it by searching my name, Marcus Guevara, or by visiting my website ThirstingforTruth.com. What you'll find out, if you watch the video, is that I did not grow up with a strong faith. I had a very powerful encounter with God at a college retreat for the first time where I met— who is now—my bride. The life I lived before I met her is not the kind of life I would want for my kids. I was driven by lust and made many mistakes. She had very high standards for herself and for the relationship. She was also strong willed and challenged me to meet those standards. I think, apart from the circumstances of that time, I wouldn't have wanted any part in that sort of relationship. However, at that time I was looking for something to turn my life around. My lustful lifestyle and bad decisions had created a void in my life that became overwhelming during my college years. My dad, who was not a very religious person, had gone on a retreat a few years prior and came back very weird. He started saying sorry more often and praying and talking about God. I distinctly remember him telling us that he wanted to be a better person. That stuck with me and years later, while I was feeling weighed down by the heavy load of guilt and shame, I decided to attend the same retreat in hopes that I would find what I was missing in my life. Well, I did, and her name was Dianne. There was something very different about Dianne. She had this amazing peace and serenity to her and I felt like my life was chaotic. I had this great void I didn't know how to fill so I figured I could use a little of what she had (or maybe a lot). Dianne made me read a book before we could start dating that was based on something called the "Theology of the Body". I'll talk more about that later but let's just say that this Theology of the Body set me on fire for learning more about my faith and very soon I was neck-deep in Theology and Philosophy books. I couldn't get enough. I was reading books while walking to class, while I was sitting in class (not good, I know), late at night, and basically every opportunity I had. I spent years studying and learning while I went through a personal transformation. All the while, I obtained my bachelor's degree in Computer Science and a master's degree in Cybersecurity, got married, worked as a software developer, joined the military, and moved into the cyber workforce.

Years later, I had become involved in an after-school program where I was teaching high school students computer science, and cybersecurity, and I thought it would be neat to turn my lessons into online courses. I remember one day as I was contemplating how I could explain some computer concepts, I realized I kept having to use analogies to explain them. I further realized that the analogies I used were either very similar or exactly like analogies I often used to explain theological concepts. The more that I contemplated, the more I began to recognize parallels between computer science and theology. I realized that as I began to apply technological analogies to theological concepts, they would not only fit but they would illuminate the topic for me. Vice versa, I was able to apply my understanding of theology and philosophy in a way that illuminated computer science. What I found was fascinating and I believe it has enormous implications for people of faith as well as those who understand technology and believe that faith is incompatible with reason. In this book, I will show you that technology is not only good but also that it is theological. It has the ability to reveal God to us. That sounds crazy, I know, but stay with me and I'll show you just how that's possible. I'll show you exactly how I hacked theology to find God in the midst of ones and zeros.

Before I go any further, I need to make a disclaimer about the title of this book. If the term "Hacking Theology" has you worried, you can relax. I'm actually using the word in its original meaning. The term hacker has become synonymous with cybercriminal, but the truth is that's a misuse of the word. Find any knowledgeable techie (especially an old school one) and they'll tell you that there was already a term designated for criminal hacker: cracker. Crackers may or may not be highly skilled or knowledgeable but, regardless, their aim is to steal and destroy. Traditionally, hackers are people who are either very skilled or very passionate about becoming skilled in technology so that they can learn, build, improve and create. Besides criminality, the biggest difference between a hacker and a cracker is that hackers have a deep desire to understand technology, usually for the sake of using it toward a good end. Crackers usually only need to understand it to the degree that they need to in order to use it for some selfish gain. Since it's always easier and faster to destroy than it is to build, hackers often spend a great deal more time diving deep into learning about some particular technology. They don't settle for the easy answers. They search, and they read, and they wonder, and they try new things. Usually they don't stop until they figure it out and then, almost always, hackers want to share their discoveries with the community.

I consider myself hacking theology in the same way. I can't settle for easy answers. When I first started asking questions about Christianity and theology, I couldn't stand people answering with "I don't know", or "that's just the way it is". Even if they were able to provide an answer, I had to really understand so I would ask other people, read books and listen to talks on the subject. I would even listen to talks over, and over, trying to get a better understanding of whatever I was trying to figure out. I have spent, and still spend, a lot of time wondering about things that I think never cross most people's minds. I constantly ask questions and usually don't let them remain unanswered for long. In addition to my own desire to understand, I have a great desire to share my discoveries with others and help them understand. I titled this book Hacking Theology because I am both trying to show how technology reveals God to us and how theology provides us the answers that we seek in regard to technology. That's a tough task. Some of us understand technology well. Others understand theology. There are probably not too many people who understand both deeply. The greatest way we can come to better understand anything is through analogies. In other words, you can better understand something when it's related or explained through something else that you already understand. Well, I understand computers and technology and have from a very young age. Now that I have come to have a very solid understanding of theology and philosophy, I believe I have a unique perspective that will help create a bridge between the two.

This book was written to help with a couple of things. First, I would like to combat the idea that technology is ruining our culture and is a hindrance to the Christian faith. So, if you are either worried about what technology is doing to today's kids or to your own life and faith, I will show you that technology is not only compatible with your faith but can actually illuminate your understanding of the Christian faith in fascinating ways. Second, this book is written to combat the idea that belief in God, and the observance of the Christian faith, is incompatible with science and reason. I mainly focus on demonstrating this from the computer science perspective, but I think that is actually an advantage. I'll demonstrate in a later chapter that computer science is essentially mathematical logic which is essentially Christian philosophical logic. So, if you're a deeply technical person who really understands computers, programming, and logic, you won't be able to put this book down. I believe you already know way more about theology and philosophy than you realize. Theology can be much like a conglomerate of encrypted ideas. All you need are the right keys to decipher those ideas. You may actually already have the keys to many difficult concepts. My goal is to show you how to apply them. Lastly, this book was written to help teach people of faith how to communicate with technical people. Yes, you can purchase this book and hand it to your son or daughter who is in college studying computer science and beginning to question God's existence, but you need to read it first (or simultaneously). I believe one of the reasons there exists this common perception that people who understand the sciences (to include computers and technology) don't believe in God, or believe religion is nonsense, is because people of faith often do not know how to communicate and relate the faith to them. Techies and computer scientists can often feel like science doesn't have a place in faith or, worse, that the Christian faith is hostile to computers and technology. As a person of faith, when you hand this book to someone you are inviting them to learn more about your world and something that you hold close to your heart. If you have no interest in learning about technology, then you are essentially asking them to do something for you that you are unwilling to do for them. People of faith need to learn to communicate to people who love technology and people who love and understand technology need a bridge to understand faith. This book is not intended to be an introduction to computer science. The book will stay at a very high level and only, at times, will I dive into a technical discussion that is aimed at teaching a concept to a non-technical person. You don't have to worry that this book might be too much for you if you don't understand computers and technology. I will explain to you what you need to know in a way that is easy to understand. I will help you learn how to communicate to technical people, but you also need to commit to put forth the effort to learn some of the language they use. I assume that if you are highly technical and/or a person who is skeptical about faith, you are already making the effort that I am speaking of as evidenced by you reading this. If that's the case, thanks for giving this book a chance. You won't be disappointed.

I'm insanely excited about the opportunity to dive deep into this world with you. I'm now married with four kids (and another one on the way) and I've actually been doing a lot of this writing late at night after everyone goes to sleep. I guess some things don't change much. The only difference between now and when I was eleven is that it's now my wife coming into the room and telling me to go sleep. I still have the same passion for technology that I had at a very young age and I've spent the last ten years developing a deep passion for understanding theology (and putting it into practice). For a while I was torn between these two worlds. At times, I wanted to leave the tech industry to pursue theology and ministry full-time. Other times, I felt like I couldn't be happier working in the cyber field. Once I began developing the ideas in this book, I realized that I didn't have to let go of one for the sake of the other. This book is a result of my journey to bring these two worlds together. My hope is that people smarter than me will take these ideas and run with them. True hackers know that amazing creations come about in enthusiastic communities. Everyone is created with the desire to seek truth and no one finds it alone, so I hope you find as much enthusiasm for these ideas as I do. I am a lover of truth. In fact, I thirst for truth. My life is and will always be driven by questions about life's greatest mysteries. Hacking is about understanding and creating. It's about deep understanding which comes as a result of tireless efforts to seek out answers to the mysteries that fascinate us. I hacked theology because its mysteries were fascinating to me and I wrote this book because I want to share what I found with everyone.

I don't know why you picked up this book. Maybe you're puzzled by others who are nuts about technology or maybe you can't get enough yourself. Maybe you're just curious to see if it is possible that technology could reveal God. Maybe you've felt a stirring in your heart—a desire to know if this world has meaning. Maybe you've sensed that technology has that answer. Whatever your reason for picking up this book, I challenge you to look deeper. Open your mind and don't stop reading. We're about to dive deep, ask difficult questions, and not settle for the easy answers. We're about to hack theology.

I hope you're ready.

WHAT EXACTLY IS TECHNOLOGY?

Imagine your job is to study people as they interact with technology. You sit down in a room with white walls, holding your coffee in one hand and a pen in the other. Your subjects sit down in front of you. Each one sits at a different station. The first station has a TV with a collection of movies for the subject to choose from. The second station hosts a popular video game. The third station is at a powerful computer and the last station contains a smartphone. Because you have a tremendous sense of humor, you decide to play the song "Every Breath You Take", by The Police, over the room's speakers. For the millennials who may be reading this, it's funny because the chorus is: "Every breath you take, every move you make, I'll be watching you...". You maintain a very serious face while they look to you for a laugh, or a smile. It never comes. The song ends and now it is time to begin. Each participant begins to interact with the piece of technology in front of them. You can't see their screens or hear what they hear. Your only data comes from watching their body language and facial expressions. You observe as each participant goes through various emotions and reactions. At times, they express joy and excitement, frustration, anger and sadness. Other times they sit emotionless, their eyes peeled open while they stare intently, concentrating, on whatever is in front of them. Hours go by and they continue to remain engaged.

What's going through your mind? What questions are you asking yourself and what are you learning about your participants? I bet that if you think about it, at some point, you have probably already made someone an unsuspecting subject of this experiment. It's not uncommon for us to sometimes watch others as they interact with technology and wonder what they are doing or why they are so engaged. Maybe you've never been into video games and you watch gamers and think, "What do they see in that stuff?" Maybe your thirty-year-old husband loves to play video games and you wonder, "How can he spend so much time playing those darn games?" Have you ever looked at someone who has been glued to their cell phone for hours and thought "What have they been doing this whole time?" or "Why are they stuck on their phone instead of watching their kids?" Maybe you haven't ever thought any of these things, but I sure have. In fact, I've asked many of these types of questions of myself. I've often made myself the subject of my own study and wondered why I had such a deep affection for technology. Although the examples I used were negative, this curiosity we all have doesn't come from a place of suspicion or judgement (though it can end up there). It comes from a desire to understand what is going on. How did technology manage to capture our hearts and minds so powerfully? It was a process that occurred over time. So, to understand we need to briefly look at the history of our relationship with computers and technology to see if we can figure out when we all got hooked. While modern technology is relatively new, it's old enough now that many of us have grown up with it as an integral part of our lives. I was born in the late eighties so most of my childhood spanned the nineties. The Nintendo, and then the Super Nintendo, were the first video game consoles I grew up playing. We got our first PC when I was in sixth grade. Since then, technology has been a big part of my life. Of course, I never really looked at technology critically until I got older. When we're young, we normally just accept whatever's in front of us. If someone hands us a candy, we eat the candy. We don't stop to ask, "What's in this?" or "Is this potentially bad for my teeth?" As we get older, we start to realize the value of questioning and analyzing things that we consume. Two important questions to ask when analyzing anything are "what is it?" and "where did it come from?" So, the first question we need to ask when analyzing the history of our relationship with technology, is, "what exactly is technology?"

I'm a lover of words and I believe looking at the etymology (origin) and development of the words that we use usually points us in the right direction. The word technology comes from two Greek words: "tekhne", which means art, or craft, and "logia" which means "a speaking discourse"[2]. However, the Greek "tekhnologia" is defined as a "systematic treatment", which sounds nothing like the previous two definitions and it doesn't seem to describe the way we use technology today either. If you look closer, however, it's easy to understand how all these things are related. Historians believe that an explosion in innovation around 70,000 years ago may have been caused by the development of language, which according to one historian, explains "why humans all over the world suddenly began to use symbolic representation and act in creative ways at the same time"[1]. Interestingly, the "technology" at this point became dramatically different. Before this time tools were strictly functional. They were used to perform a very specific function for survival. When innovation exploded (potentially due to the advent of language) humans began to become creative, expressive, and their technology performed functions that were not exactly necessary for physical survival. They became interested in exploring beauty and meaning through the use of their tools. This sort of exploration of beauty and meaning through art often leads to the creation of more functional technologies as well. Humans quickly advanced after this point and large civilizations soon found their way into existence. We'll come back to that later. For now, let's focus on the relationship between technology and beauty. Another word we derive from the Greek "tekhne" is the English word, "technical" which we normally use to describe proficiency or accuracy in some art or craft. Boxing is often described as a technical art. Painters, and martial artists, develop a particular technique in their craft. It seems today that most of the American population has become experts in the art of the English language since so many people often use the word "technically" to correct someone or to further clarify a point (I only do it if I need to make myself sound smarter than the other person. That's a joke, of course). People who play basketball are constantly trying to improve their skill to become more accurate in shooting and more proficient in dribbling. Artists aim to become more skilled sketchers and painters so they can more accurately represent their visions or the objects they are trying to replicate. The process of individuals becoming better at a particular trade usually leads to an overall betterment of the trade itself. Think about how the skills of individual football and basketball players have improved over the last 60 years. Individual players work hard and significantly improve their skills to become more competitive. In turn, others are forced to work harder to stay competitive or risk becoming outmatched. As a result, and over the course of 60 years, the average height, muscle mass, speed, agility, and endurance of professional basketball players has greatly increased. However, sheer determination or work ethic are not the sole reasons for this evolution. Innovation has played a big part. Players learned ways to train more efficiently and effectively. Weight lifting has evolved to include a variety of equipment that helped players target more specific muscles. Other training inventions have helped promote repetition of particular skills. For example, at some point, someone invented a curved plastic accessory called the basketball return, which, when hooked onto the bottom of the hoop, guides the basketball back to the shooter. This device helps to eliminate the need for the shooter to run after the ball after each made shot. The practicing shooter, then, should be able to take more shots in less time and, thus, become a more accurate shooter. This kind of evolution is prevalent in every kind of art, craft, and trade. Artists have evolved from using pigment and rock to drawing on digital tablets and designing three-dimensional cartoons. In between, there have been innovations in the tools that artists use such as oil paint, brushes, pencils, ink pens, digital styli, and computer programs. These kind of devices (to include the basketball return) are considered technology and they are born out of the technical exploration of beauty and meaning. Just as an artist finds meaning in expressing his or her ideas on paper, sports players find beauty and purpose in competition. For many, their art, or craft, becomes a way of life. As they become more skilled, their skill becomes more valuable in society. People find great entertainment in watching others demonstrate their high level of skill. The NFL (National Football League) is worth billions of dollars and the Super Bowl commands millions of viewers each year. The Hollywood movie industry, likewise, has a constant churn of new movies each year that rake in hundreds of millions of dollars. As people recognize the value that society places on highly skilled athletes, and artists, they begin finding ways to monetize their skills. They erect large venues and charge large sums of money for ticketed events. They contract television deals and charge audiences for live broadcasts. In order to conduct these sort of business arrangements they need to, of course, employ many people to carry out various tasks. As people find success, others try to replicate that success. Others look to follow in the footsteps of masters, and so books are written, courses are developed, and universities are erected to systemize and deliver on the promise of helping others achieve success in their trade or craft. Innovative technologies find their way onto the scene helping to maximize profits, decrease safety hazards, and provide a better experience. As a result, you have extremely structured and systematic, or technical, approaches and solutions to the deep exploration of art and craft. Think, for just a second, of how amazing it is that people are so fascinated by some expressions of art (including sports as a form of art entertainment) that they become million, and even, billion-dollar industries. In a profound way, art speaks to us and brings meaning to our lives. Literally, music is one of the arts where speaking is a large part of the form. Even when language is not involved, we communicate ideas and emotions through art that are sometimes hard to put into words. This is an idea that we will dive deep into later. Our first task was simply to define the word "technology" as we plan to use it in this book. So far, we know that technology involves a systematic approach to an art or craft but how exactly does that describe what we normally refer to today? Well, if you paid close attention you might have noticed that technology itself was never the object of our focus. Even as we explore the question, "what is technology?" we have to look at the purpose that it serves which is to assist us in some way. Technology is never an ultimate end. It's something that provides assistance to an end. In many cases, it assists us in solving problems to bring about the end that we seek. Therefore, technology is really anything that we create to assist us in achieving something we may not be able to achieve otherwise.

It makes sense then, that nowadays technology is usually used to describe computers, smartphones, and smart devices. Computing devices process and translate information and rely on accuracy to work according to their design. You wouldn't want to dial 911 in an emergency only to reach 411 because of inaccurate processing on the device. Thus, they are technical devices. However, computers are not designed to simply process information accurately. They are also elegant devices. Often, the hardware itself is a form of art. They provide entertainment and they assist us in the craft, or discipline, of finding amazing solutions to difficult problems but they are also objects of beauty. Modern computers certainly fit this description. However, they are obviously not the only type of technology according to this definition. Why is it, then, that computers (to include laptops, tablets, and phones) are the poster boy for technology these days? This brings us to the second question we should ask in analyzing our relationship with technology which is, "Where did it come from?"

INVENTIONS THAT TECH'D THE WORLD

If Madonna were a millennial, she probably would be singing "I am a tech'd out girl, living in a tech'd out world". That's a dumb joke, I know. Still, it does seem that technology is everywhere and is only growing more popular. But how exactly did our world become so "tech'd"? The long answer to that question is a book on its own. You could probably design an entire college course on the topic. For now, we're going to focus on some of the most notable inventions that tech'd the world. Because I believe that we are a part of a technological evolution that started a long time ago, I'm gonna have to bring up some ancient history. My wife hates when I do that during fights, but I think it's necessary in this case. Since we've defined technology as something which assists us in achieving what we might not otherwise be able to achieve on our own (be it for art or function), we're going to take a look at some of the most notable achievements in history and the tech behind them.

Before the agricultural revolution, early people survived solely on wild food and wild animals. We call these the 'hunting and gathering' societies. Their lifestyle required that they be constantly on the move; dependent on the location and abundance of plants and animals. Around 10,000 BC humans began to develop agriculture through the domestication of plants and animals. The development of agriculture radically changed the lifestyle of previous hunter-gatherers as they began to settle and live off the farmland instead of wandering in search of food. This created other dependencies, though. For example, farmers then became dependent on rain to water the crops. Although farmers could support many more people through farming, a drought, or diseased crops, could mean devastation for the community. By 5,000 BC, humans had figured out how to solve one of those problems by building canals to channel water from rivers, and lakes, to their crops. This allowed them to become less dependent on rain and ensure their crops were a more reliable source of food all year. Farming eventually replaced hunting entirely in some cultures because of its ability to support large numbers of people with a smaller footprint. Historian Daniel Headrick questions whether this should be considered progress as archaeologists have discovered that early farmers were shorter, more poorly nourished, and more disease ridden than their hunter-gatherer ancestors[1]. Regardless, people took to farming and figuring out ways to improve their methods. When a new problem would come about, people would figure out a new solution to resolve it. For example, in Mesopotamia, the Nile flooded its valley every summer and fall. In response, the Egyptians developed a technology, called nilometers, to help them predict when flooding would occur. Agricultural innovations provided relief to the persistent fear of famine experienced by these early people and allowed the population to grow. The surplus of food and greater access to water also allowed people to create settlements which, over time, gave rise to cities and civilizations. Eventually, people figured out that water could be a source of great mechanical power and developed watermills and other water propulsion systems— like the Vitruvian waterwheel—to make their lives easier and increase production. Soon, civilizations found themselves with a new problem. Farm land and access to water became extremely valuable and land ownership became a point of contention and war. Civilizations found themselves vulnerable to attack from those who desired their property. In response, civilizations began to produce full-time warriors and began to develop innovative defenses to protect their territory. In turn, newer innovations came about to counter improved defenses. Advances in weaponry spurred advances in armor, which then produced the need for more powerful weaponry. We'll continue to see this cycle of new problems inspiring new solutions which inevitably bring about new problems again. Since we are still on agriculture, we'll come back to the topic of war and weaponry in a minute. Surpluses of food also allowed people to spend time learning and perfecting new ways to contribute to society. Some examples include pottery, weaving new types of clothing and making metal objects through mining and smelting. The production of iron brought about the iron age which radically impacted the world in multiple ways. For one, it allowed peasants access to better tools. Blacksmiths could make iron products for much cheaper and so farmers, cooks, carpenters and even blacksmiths began using iron tools. New iron tools allowed farmers to perform their work much more efficiently and, like the result of food surpluses, this also allowed families time for other activities like making salt, sewing clothes, and crafting luxuries such as jewelry[3]. However, the more significant change brought about by iron production was that it changed the dynamic of war, allowing newer civilizations to topple older and traditionally more powerful civilizations, giving birth to new empires. The idea that it improved the lives of many people but also contributed to the death of many others is observed by philosopher Pliny the Elder in the following quote:

> We will now consider iron, the most
> precious and at the same time the worst
> metal for mankind. By its help we cleave
> the earth, establish tree-nurseries, fell trees,
> remove the useless parts from vines and
> force them to rejuvenate annually, build
> houses, hew stone and so forth. But this
> metal serves also for war, murder and
> robbery; and not only at close quarters,
> man to man, but also by projection and
> flight; for it can be hurled either by ballistic
> machines, or by the strength of human
> arms or even in the form of arrows. And
> this I consider to be the most blameworthy
> product of the human mind[1].

It's important to recognize that iron production was considered a technology. At the time of its founding, it was literally "cutting edge". This phrase conveys perfectly the imagery that technology can be very divisive. This is an idea that we will expand upon later in the book but it's an idea that you can highlight in your mind for now so that you recognize the pattern of technology bringing, at once, the potential for great good and great evil. During the Han dynasty, we also see the production of iron help bring about the first known instance of mass production. Iron has since remained an essential element and was part of the building blocks in the development of steel which is currently the most important building material in the world[3].

Another one of the most prolific technological innovations during the advent of civilization was written communication. The first system, called cuneiform, was developed by the Sumerians and consisted of a clay tablet where figures and images where etched or engraved. It was originally used for things like making lists and keeping track of temple donations[1]. Writing has changed dramatically through the ages and eventually evolved to using numbers and phonetic symbols as we have now. What is most interesting about the evolution of writing is that, at some point, it was new and no one had ever been able to accurately document a thought outside their minds before. At least not through language. In the earliest hieroglyphics, people could use images to help document their thoughts and stories, but images cannot be interpreted   as   accurately   as   written communication. When phonetic languages, and symbolic letters, became used for writing people could then more completely convey their thoughts to each other. We still take this sort of thing for granted. I find it amazing that when I pick up a book that was written by someone fifty, one hundred, or one thousand years ago, it's almost like I can go back in time and talk to that person. Written communication allows ideas and knowledge to be passed on from generation to generation. However, written communication would be rare and not very far reaching without the invention of paper. Paper is attributed as being invented in China around 105 AD (although there's evidence it may have been around two hundred years earlier) and is obviously still in use today. Paper stimulated the production of written works and would eventually make books and scripts more accessible which eventually helped to promote literacy among the general public. It's common knowledge that paper was, and is still, often used for artistic purposes as well.

As mentioned earlier, war is also a very big driver of technological inventions. New technologies often change warfare and give one side a significant advantage over the other. Historically, the shift in advantage usually goes back and forth between the best defense and the best offense. Defensive innovation produced high defensive walls and armored knights. Offensive innovation produced catapults and battering rams. Headrick explains that, unlike agriculture, military technology diffuses rapidly from one people to another. This is because warriors can imitate their enemies without the need to replicate their enemies' specific environment (like they would with agricultural techniques)[1]. Because of this fact, when warfare became rampant after the fall of the Roman empire, the advantage never stayed on one side for too long. A theme we are developing here in this chapter is that the history of technology, like all history, repeats itself. Currently, in cyberspace we have the issue that defensive and offensive tools and techniques change so rapidly that the advantage often shifts between attackers and defenders. Every time some new exploit or offensive technique is released or discovered, there is a rush to build a system or software that detects and prevents the attack. Utilizing modern intelligence techniques and offensive cybersecurity skills, defenders have found ways to attribute attacks, uncover persistent footholds, and eradicate threats. However, as soon as enemy techniques are discovered and accounted for, attackers quickly devise new, and extremely innovative, techniques that allow them to gain new concealed footholds. I firmly believe, at some point someone will get a clear and consistent advantage in the cyber realm but only through a groundbreaking advancement that will be the product of consistent innovation and adaptation. However, I also believe security will always be an issue as long as computers are in use. I am confident in this because I believe our issue is not unlike the past when the arrow stifled the swordsman from a distance; when wooden arrows then became useless against the armored knight; and then when the iron crossbow pierced through the knight's armor with deadly force. While technological advances did shift the advantage more quickly, there were times when the advantage was held for long periods, ironically, due to a technological advancement. The stirrup, for example, was a simple but very pivotal invention that provided lance-bearing knights with the decided warfighting advantage for several centuries. Like any other part of life, technological advances may change a dynamic temporarily, but innovation must be a process of adapting, learning, and re-inventing. What works now will always be bested by the next advancement. Just ask the lance-bearing knights if you happen to see any around.

So far, we have taken a very high-level look at agriculture, tool production, the development of communication through paper and writing, and inventions that were spurred by war. There are so many important inventions in early history that impacted the world, but our task isn't to explore every significant invention. We set out to ask the question, "how did our world get so tech'd?" The reason I covered these categories at a high level was not so much to show the significance of the inventions themselves but to show where they come from and why that's so important. One of my goals is to help you focus on how the desperate fight for survival, the desire to make tasks easier, the creative exploration of communicating one's thoughts and ideas, and war, bring about the greatest innovations. Inventions in survival are pretty easy to understand. Famine, death, war, and desperation provide great motivation to problem solvers. Great discomfort and the threat of death make everything else seem far less important. Large numbers of people can more easily find the motivation to dedicate their time and effort and work together to evolve and survive. This is the reason that, throughout history, many of the greatest inventions have come about during times of war. However, when people are not under some serious impending threat of death or illness, they are likely to be spending their time on leisure or creative efforts. When a community or society's survival is not in question, they tend to innovate in the areas of curiosity, exploration, task improvement and easing life's difficulties. I include the accumulation of wealth and luxury as an aspect of easing life's difficulties since, wealth allows one to decrease the amount of labor they must perform, and luxuries are often satisfactions which rid one of the hungers or absences they experience. War can be an aspect of survival for those who are under attack but for the attackers themselves, war often carries the objective of the accumulation of power and wealth. On a smaller scale, criminals will often work extremely hard to find ways to steal, kill, and destroy. Therefore, some obviously find their motivation for innovation in evil rather than good. Aside from life or death situations like famine and war, the desire to innovate often begins with curiosity. Human curiosity then leads to contemplation and eventually experimentation. We like to see if our ideas work and if they will allow us to produce more fruit (literally and figuratively). If we can produce more, we usually then want to see if we can we produce something better or more efficiently. This leads to the innovation, and production, of tools. As we demonstrated in Chapter Two, once we have mastered functionality in some area, we move to greater exploration and creativity. This process is never ending as war and chaos are constant disruptors of peace and progress. People tend to notice where others are successful and plentiful. Out of jealousy, envy, or starvation and desperation, they wage war to reap the rewards of others' innovations. Sometimes natural chaos ruins cities, civilizations, and progress. People then need to rebuild and reinvent to not only recover, or continue progress, but to now anticipate, and defend themselves from, future enemies and natural disasters. Although it may seem, to us, that technology progressed at turtle speed in the beginning, this process started from the moment of human creation and has been cycling ever since. So, the loose answer to the question of how our world became so tech'd is that it's a natural progression of life. We have to invent, and then we want to invent, so we do. I'm not sure that there is a way to avoid technological innovation since humanity might not exist now if we hadn't increased the food supply, found cures to various plagues, learned how to adapt to new environments and learned how to survive and recover from natural disasters. I also don't know that anyone can stop dreamers and explorers from risking their lives and sacrificing sleep in order to create and discover. It's not like there haven't been efforts throughout history to do just that. We innovate because it is our natural response to problems and mysteries. Technological innovations increase in sophistication through the normal process of evolution described above. Solutions to old problems encounter, or create, new problems which eventually entice problem solvers to look for new solutions and so goes the cycle. However, if we fast forward through ancient history and into more recent times, we find a more precise origin to our current tech culture in the industrial revolution (though the industrial revolution occurred over a few centuries). Remember that our purpose is not just to create a timeline but to see if we can gain some insight into the question we asked at the end of Chapter Two. In that chapter we answered the question, "what is technology?" and ended the chapter with the question, "where did technology come from?" In this chapter, we have been slowly answering that question. However, we're not looking to simply create a historical account. If you remember, these two questions came about in an attempt to answer another question, which was: how did technology manage to capture our hearts and minds so powerfully and when exactly did we get hooked? In that context, we are looking to see if history can help us better understand how our lives have seemingly become so dominated by technology. We're still trying to analyze our relationship with technology at this point. Keeping all this in mind, let's take a high-level look at the industrial revolution.

> *Before the advent of the Industrial*
> *Revolution, most people resided in small,*
> *rural communities where their daily*
> *existences revolved around farming. Life*
> *for the average person was difficult, as*
> *incomes were meager, and*
> *malnourishment and disease were*
> *common. People produced the bulk of their*
> *own food, clothing, furniture and tools.*
> *Most manufacturing was done in homes or*
> *small, rural shops, using hand tools or*
> *simple machines* [9].

Amazing, isn't it? This is a deep contrast from how many of us live now on basically every point. Much of the world is now made up of big cities. Even small cities with a population of twenty to forty thousand are big when compared to old rural communities. Extremely few people in developed countries have a daily life that revolves around farming and the average person has the ability to attend college, get a good paying job, and afford a more-than decent home and vehicle. Virtually no one produces their own food, clothing, or home items. I mean out of necessity, or as a complete lifestyle, not for hobby or unnecessary supplement. Lifestyles have changed dramatically in these areas since the industrial revolution which helps to demonstrate just how big of an impact industrialization had on society. With such a big subject, we have to be careful to really only concentrate on some of the aspects of the industrial revolution that led us to our current marriage with technology. There are a few major aspects that we will focus on. The first, is that industrialization greatly increased the material wealth of society and changed the way people lived. The average person is in a much better financial position than before industrialization. We live in a time now where a good idea can—with enough hard work and maybe some luck—be the start of a new, and lucrative, career. It's hard to understand that there was a time when this was not the case. Wealth was based on status and family lineage. People's financial situations rarely changed since their family and status did not change. Increased wealth among the average person has far reaching impacts in terms of technological innovation. Industrialization brought about mass production of food. Because larger sets of people did not have to rely on manual labor (ex: farming) for sustenance or income, they could then apply their creativity and critical thinking to innovation. Another major aspect of industrialization was the way in which it changed the average person's lifestyle. As innovations became more common and the general wealth of the average person increased, there was higher throughput in the exchange of goods and services. People spending more on minor transactions for things that they wanted, but did not necessarily need, created a market for businesses who sold low to medium cost items. The industrial revolution made this possible through factories of mass production which created materials and products in a fraction of the normal time and cost. Thousands of people flocked to industrialized cities in search of employment and cities that boomed with factories and industrial progress quickly became overcrowded. For the first time in history, people began to work outside of the local area of their homes[10]. So, the industrial revolution provided us the introduction to the daily work commute. You can thank the industrial revolution for this book since I spend a great deal of my time researching, and writing, on the commuter bus each morning and afternoon. At this point, you might feel like you are starting to see the roots of our current lifestyles and technologies. As cities became densely populated, forcing people like me to commute to work, they also became ridden with crime.

Overcrowding created social dysfunction and forced the innovation of an organization of full-time government employees who were trained in innovative methods to police the people, prevent crime, enforce law, investigate criminal acts and catch criminals. All of the modern local and federal police forces and investigative agencies have their origins in the industrial revolution. It's hard to qualify police forces and investigative agencies themselves as technology but the techniques and tools they use certainly involve a great deal of advanced technologies (relatively advanced in each time period). Much of the sophistication in the tools used by police forces and investigative agencies were initially spurred by the great need in overcrowded cities that were overcrowded and dense as a result of the industrial revolution.

Another interesting correlation comes from the change in the work day. If you think that working between 8 am and 5 pm Monday through Friday is normal, well, you're right. It's normal now. But it became normalized during the industrial revolution. Before the industrial revolution people relied on the sun for light and were subject to the weather. Therefore, seasons determined the work day. Farmers can only work when there is light outside, so they worked from sunrise to sunset. They had a tough work day and they didn't have weekends off to recover. However, during the winter months, apparently, the lack of electricity and no central heating kept people in bed ten to twelve hours a day[10]. While that might sound awesome, I think I prefer central heating and air conditioning. The advent of central heating and air conditioning allowed work to be performed indoors regardless of weather conditions. However, new problems emerged as working conditions in the industrial revolution were far from ideal. People worked long shifts, averaging fourteen to sixteen hours a day with no recovery periods during the winter. Women and children had the worst circumstances as they were given the most dangerous and grueling tasks. Before, people had jobs that were based in families and communities, they enjoyed certain benefits like not worrying about being fired or laid off. They worked with people who were a part of their family and cared for their well-being. Now, people began to work for large companies run by people they didn't know. Neglectful working conditions sparked revolutions for better treatment. While there were great companies then, and many great companies now, to this day some of the issues of working for a company continue to be a point of contention for workers. People still complain about being overworked and feeling as though they are treated like a number instead of a person. People still fear being suddenly without a job and losing their livelihood due to layoffs. It's not like this fear didn't exist before the industrial revolution. Loss of crops to hungry animals, or a drought, could cause people to lose their livelihood for some time. Some probably never fully recovered during severe incidents. However, the point is that our modern experience of working for a large company with set hours, indoors, under a complex contract that can lead to lawsuits over rights, discrimination, etc., is very much a product of the industrial revolution.

Of course, the advent of companies and innovators is not all bad. Since industrialization brought more wealth to the average person, made tools and materials more available, and gave people more time to experiment, the late 1800s and early 1900s brought a wave of amazing inventions. In 1712, Thomas Newcomen developed the first steam engine to pump water out of mines. A few decades later, James Watt improved the steam engine which would then go on to become the source of power for ships and trains and radically changed the face of travel and trade. According to Headrick, the steam engine was the most visible symbol of the technological progress during the industrial revolution. Further, he notes that nineteenth century Europeans and Americans equated technical progress with progress in values or the "rise of civilization"[1]. He quotes William Huskisson who believed that the steam engine was a powerful tool that assisted in the moral obligation to further civilization. This was not a new idea as I'm sure that agricultural inventions were also viewed as moral necessities since they relieved famine and allowed the population to grow. However, the idea that technological innovation was a moral good and a pathway to success became something of an obsession during this time as we see rapid births of new inventions by the lower class as well as the elite. In the 1850s Henry Bessemer figured out how to mass produce steel. Because of mass production, iron and steel became inexpensive enough to become the popular material among tools, machines, buildings, and eventually personal vehicles and home appliances[9]. The industrial revolution provided a platform where people left behind agricultural work and moved into cities and communities where innovation, engineering, mass production, profit, exploration, and discovery through technology now became accessible to the average person. Having access to inexpensive materials and tools allowed people to build upon other inventions. I have to control myself here because there are so many awesome and important inventions during this time, I'm struggling to not bring up cool inventions just because they were cool. I'm going to try my best to stick to inventions that have relevance to the evolution into our current technology.

Earlier we discussed four different categories where the greatest technological advancements normally occurred: survival, the betterment of life and easing life's difficulties, exploration and creativity, and war. The industrial revolution ushered in amazing inventions in transportation like large ships and airplanes; it brought about terrifying weapons like the machine gun and the atomic bomb; it changed transportation through the combustion engine and the consumer automobile; it made seemingly impossible exploration a reality through rockets launched into space; and it changed our daily lives through tons of smaller inventions like the household vacuum. However, in regard to our current technology, communication is the area from where modern technology finds its birth. I placed communication under the category of exploration and creativity. However, like all the inventions we discuss, communication innovations can fall under any of the categories at different times. The reason I primarily classify it as a result of exploration and creativity is because there is an inherent desire in humanity to express their inner thoughts and feelings more accurately. We need to be understood by those closest to us, but we also want to communicate as far and as wide as possible. Nowadays, we have YouTube, podcasts, radio, live television, live concerts and speeches with thousands of people in attendance, instant email, blogs, and the list goes on. Innovations in communication did not start in the industrial revolution, but great strides were made that projected us to where we are today. It was during this time that scientists figured out that they could send messages over electrical impulses and the telegraph was invented. Samuel Morse developed his "Morse code", which assigned each letter of the alphabet to a pattern of dots and dashes. The first message sent by telegraph in the US using Morse code was "What hath God wrought". The phrase taken from the Bible (Numbers 23:23), where a man named Balaam is sent to deliver a message from God. Apparently, the phrase was suggested to him but it's a very fitting question as the telegraph would help alter long distance communication forever. In 1866, North America and Europe were connected by telegraph cable. Shortly after, India was connected as well. The world was already becoming networked for communication through cable before the 1900s! The telephone was developed through various experiments between the mid to late 1800s. Famously, in 1876 Alexander Graham Bell was awarded the US patent for the telephone. With this invention, people could then call and hear each other's messages instead of having to translate Morse code. The desire to communicate continued to evolve and eventually we learned that we could send signals and sounds wirelessly. In 1899, Guglielmo Marconi sent the first wireless Morse-code transmission across the English Channel. Inventions in radio communication then allowed voice and sound to be carried wirelessly over long distances. AM Radio exploded in the 1920s. Broadcasters began to use the platform to deliver instant news and entertainment. Because of mass production, the common person could afford a receiver and thus be in the know of current events as they happened, listen to trending music, follow sports games that they were not be able to attend in person, and catch mystery or drama shows to help pass the time. What's truly amazing, is that apart from the cost of the receiver people were able to listen to these radio broadcasts for free! Broadcasters figured out that they could operate radio at no cost to the consumer because companies would pay a ton to advertise their products to the extremely large number of consumers who became instantly hooked on radio. If this doesn't sound familiar to you, you may be living in a different century. Google has built a billion-dollar business off of free services that are subsidized by advertising. Of course, Google isn't the only company that inundates their consumers with advertisement in order to provide free, or cheaper, service. They are just the easiest example. The advent of radio mobilized public communication and product advertisement. People could now take entertainment nearly anywhere with the trade-off being that companies also essentially followed them around pitching new products and services. While definitely annoying at times, this isn't always a bad thing. I'm actually appreciative of having certain products and services available to me at little, or no, cost in exchange for advertisement. It's usually only annoying when I am pitched products I don't want or care about. I don't mind seeing trailers for movies I want to see or advertisements for electronics when I'm in the market to make a purchase. I'm getting off topic here. The point is that radio helped set trends in business and lifestyle that continue to this day.

As mentioned, innovations in communication are really where our "tech'd" culture finds its birth. Later inventions like cell phones, wireless internet, streaming entertainment, social media and more are descendants of innovations in communication. However, our modern lifestyle and how we use these inventions were also largely influenced by other areas of the industrial revolution. Another category of innovations which I described as easing life's difficulties has had significant impact on how we live our lives today as well. While some inventions in farming, like water canals, were spurred by necessity and survival, other inventions came about because farming is hard, and people desired to make it easier. During the industrial revolution, there were a wave of inventions that simply eased the difficulties of life or promised to make it more pleasurable. The household vacuum was mentioned earlier as an example. Other examples include better, and more options in, clothing, indoor air conditioning and heating, electricity inside the home (for lighting and cooking), and the refrigerator. Even the modern household bathroom is a product of technological innovation during the industrial revolution. I thank

God every day for the person who invented the toilet. Well, maybe not every day but I can't imagine waking up in the middle of the night to walk outside during freezing temperatures and, uh, "take care of business". The kind of business that doesn't make you any money but that you have to take care of anyway. You know what I'm talking about. These are obviously only a few of an extremely large number of innovations in this category. The most astounding aspect of these inventions that made this time period different than earlier civilizations is that they quickly became accessible to nearly everyone in industrialized societies (not only the rich). Like with free radio, this accessibility came with a blessing and a curse. Daniel Headrick points out that industrialization brought about mass production and mass production required mass consumption. Many innovations were affordable and accessible to the average consumer because they were produced in large amounts. As we know, this creates a cycle where people usually more easily grow tired of what they own and since they can afford to replace their items, they do. This is where we see the beginnings of the "throw away culture". In turn, companies eventually realize they must churn out more products, or new and more competitive products, to continue to maintain or increase profits. This helped to create the need, and opportunity, for mass advertisement. As explained in the previous paragraph, radio became almost an overnight sensation because manufacturers were able to mass produce receivers at low cost. However, radio would have been a failed venture without mass consumption. Mass consumption in radio changed the lifestyle of many in society in how they experienced news and entertainment. Mass consumption of products that were aimed to ease life's difficulties changed the lifestyle of many in how they functioned at home and at work, how they ate, and how they treated the materials that they owned. Mass consumption made products more accessible, made life easier in certain areas, but also brought with it new problems that people never had to deal with before. Indoor heating and air conditioning makes it easier to feel comfortable indoors but good HVAC systems can be expensive to install and maintain. When a compressor dies, or the system needs to be replaced, the consumer finds himself with an unexpected—and likely very large—bill with potentially no other options for heat or cooling. During prior periods where people managed their own homes, they often managed their own survival through building their own heating mechanisms and cooking methods. In an industrial society where people began to work in factories and specialize in a particular trade, they relied on others who specialized in things they needed. So, if someone's air conditioner broke, there is a good chance they had no idea how to fix it. Again, this should sound very familiar as this is true of many people today. While it is true that people have always relied on each other in society, the industrial world became increasingly reliant on tools and specialties that didn't necessarily help a person build basic survival skills. For example, large factories produced many people who became highly specialized in a particular area of an assembly line—developing a skill that they likely would never use outside of that factory. Because mass consumption also meant that when people grew tired of their materials (clothes, decorations, toys) they could afford to replace them, they more readily threw things in the trash. Mass distribution of packaged goods, foods, and the increasing habit of replacing (and throwing away) items, caused industrial societies to have to figure out sophisticated systems for garbage disposal. It should now be easier to recognize that many aspects of our current culture that we feel are normal really only became common within the last two centuries. Technology has had a significant impact on the way we live and the way we think.

So, we've discussed innovations in each of the categories that I set out earlier in this chapter, but this is where things get really exciting. At a point where they merge, an innovation called the computer produced the greatest form of technology that transformed the world in such a radical way that we have moved from an industrial revolution into a technical revolution. There are many inventions throughout history that show a progression toward a systematic machine that could be programmed to compute numbers and automate tasks. From the 1820's to his death in 1871, Charles Babbage, designed and partially developed the Difference Engine and the Analytical Engine. Two machines that set out to automate complex calculations and programmable computations. As a founding member of the Royal Astronomical Society, Babbage's initial motivation was to figure out a way to automate astronomical calculations [4]. Even though Babbage ultimately failed to convince the government, and society, of its potential, his ideas were far ahead of his time and are considered the first semblance of the modern computer. In 1888, the US Census Bureau was still drowning in work from the 1880 census due to the population boom and decided to commission a contest for a more efficient way to count the 1890 census. Herman Hollerith introduced the idea of a machine that utilized punched cards which annihilated his competition and won him the contract[5]. While the prior 1880 census was still being manually counted almost a decade later, Herman Hollerith's machine was able to perform the 1890 census in a mere six weeks! His invention was a big step toward the invention of the modern computer and punched cards were in use until the 1970s to create, and edit, computer programs. Hollerith went on to start a company, The Tabulating Machine Company, which later was sold and merged into a company that was run by a man named Thomas J. Watson. That company was renamed to International Business Machines (IBM). Amazingly, IBM is still a successful business with innovative technological products in 2018. The population boom; the formation of corporations and factories; mass production and mass consumption; and the desire to innovate and create tools to ease difficulties, built the foundation for the success of companies like IBM even in the midst of the Great Depression. The corporate world was beginning to discover that there was a large and untapped market for computational devices that could solve complex problems quickly and accurately.

As it so often does, war did a great deal to speed up the development of computational devices like the ones produced by IBM. During World War II, the Germans were able to communicate across open radio waves by encrypting their messages using a complex code machine called Enigma. A counter machine was created to decipher the German messages through rapid computations of potential key combinations. Colossus could compute twenty-five thousand characters per second. The machine was almost like an anagram solver that rearranged letters at speeds that ten dozen humans couldn't come close to matching. Another notable invention during wartime was the ENIAC (Electronic Numeric Integrator and Computer) which came about in order to help compute firing tables for use in artillery during the same war. Interestingly, the work of computing firing table numbers—before the ENIAC—was done by humans whose job title was "computer".

This was not uncommon as there were many jobs in the mid-1900s, which required human computers to complete mathematical tasks. The 2016 movie, Hidden Figures, depicts the inspiring true stories of three women who worked as human computers and assisted in helping NASA to put a man into space

(Project Mercury). While the movie contains some added dramatic fiction, it shows how crucial human calculators were to the space race. Getting a man into space first became an obsessive competition between the U.S. and Russia. Projecting the precise speed, angle, distance, and weight needed to ensure the spacecraft could return as safely as it launched required constant calculation, and verification, of theories. No one had ever attempted to launch a person into space. No one knew exactly what was needed or if it was even truly possible to put a human into space orbit (safely). Getting the computations right would mean Alan Shepard would live and humanity would have kicked open the door to a whole new area of exploration. Getting the computations wrong would mean the death of a man, the loss of a major investment, and a major setback in the hope of exploring space. The sort of assurance needed to take the major leap into human space travel was helped by the IBM machines NASA purchased to provide intensely complex calculations at an unprecedented speed. The stellar results of the machines employed in this project helped change the perception of the machine computers' ability to perform the sensitive jobs of a human computer not only efficiently but accurately as well as at far greater speed. NASA and other companies began to slowly phase out human computers for machines, and, as the movie Hidden Figures depicts, people began to recognize that there was great career potential in learning how to operate these new programmable computers. The first computers were extremely expensive and so big that they took up an entire room. Eventually, inventions like the transistor reduced the size of these enormous computers and a growing market reduced the price of hardware components. The advent of personal computers and user-friendly operating systems like Microsoft Windows, and the Macintosh, put computing power, and virtual fun, into the hands of the average person. The personal computer combined with the invention of the internet began the world's journey to becoming tech'd. The internet is a child that came from an invention at the Defense Advanced Research Projects Agency (DARPA) called ARPANET. ARPANET was the first network of computers across the U.S. To extend the network, DARPA invented the Internet Protocol (TCP/IP) that allowed sub-networks to intelligently route information between connected computers anywhere in the world. It didn't take long for the internet to explode into a world-wide sensation and take over every industrialized society in the world.

I mentioned earlier that communication innovations gave birth to modern technology. The internet has been an absolute game changer in the progress of world communication. It placed an encyclopedia of information at the world's fingertips and allowed people from around the world to engage in conversation and collaboration with each other. We now transfer knowledge of history and tell stories through video, recorded audio, and printed text, which can all be easily copied and distributed around the world. Of course, the content of our communication is also much more complex now and that's because technology has helped us to communicate faster, more easily, more beautifully, more widespread, and, most importantly, more accurately. Mass distribution of books, paper, video and audio have provided greater access to information which has contributed to increases in literacy, public speakers, writers, and assisted people in communicating their ideas more effectively with family, friends, business partners and strangers. The reason this is such an important point is because communication is such an important aspect of our lives. We all have a powerful desire to express ourselves and to understand others; to know and be known. Our ability to effectively express our ideas and understand others is central to building a relationship, a marriage, a family, a community and a nation. We rely on effective communication to conduct business, and education, to avoid conflict and war, and to express love and affection. We now rely heavily on technology to facilitate our communication and to assist us in expressing ourselves more accurately. This book is a form of my leveraging technology to express my own thoughts and understanding. Technology will help me to write it faster, catch grammatical errors, send to an editor in a different part of the country and then distribute it around the world. Without technology my thoughts might not get farther than a campfire discussion or my journal (and I don't even have a journal). With technology—and no more than a laptop and access to the internet—my ideas can potentially reach and change the world. The computer and the internet opened up a world of possibilities that seems to be limited only by our imagination. This is the precise answer to how the world became tech'd. Long before we had Google Maps, there was a great deal of undiscovered land. Of course, no one knew just how much land was out there. People would look out into the ocean and wonder if the water went on forever. They wondered if there was more land out there or if one would just fall off the earth once they hit the edge. Once people began to build ships that would endure trips across the ocean, they began to find new land which then opened up a new world of possibilities. Those struggling with limited resources or looking to expand could search for answers to their problems or simply feed their appetite for discovery. At the same time, they were providing solutions to problems ailing their community, bringing pride to their country while expanding its domain, and inspiring others to expand their imagination. They brought hope to others by challenging what was believed to be impossible (or non-existent). The modern personal computer (to include mobile devices like cell phones and tablets) has been like a ship that discovered the land of the internet in a world that can be accessed from nearly any physical location. No one currently knows just how far the waters of the internet extend and what sort of things are possible in this still fairly new land. Here's what we do know. This new land is not a physical place. Rather, it's somehow the discovery of the depths of human interaction and potential.

People are now discovering new worlds and new possibilities in technology in order to connect with others, gain an edge in war; to innovate to make our lives easier, solve major problems of disease, hunger, and the threat of death, and so much more. This has led us to the point where generations have grown up inundated with technology that we rely on for all of the above. The world became tech'd because humanity saw in technology the promise of a better life. No different than those who set sail across the ocean looking for a land filled with milk and honey—a new land that held nothing but potential and an escape from the problems of the old. When technological developments began to save lives, progress society, help us discover new worlds, and connect with others on a deeper level, people began to see in technology a great hope. However, because of all the good that technology brings us we sometimes can be swept up with the idea that only technology will solve our problems. I have met people who have more faith in technology than many religious people have faith in God. Naturally, you might anticipate that this might cause some friction among those two camps. You'd be right.

Even a virtual world isn't exempt from chaos and war.

ANTI-TECHS AND THE COMING WAR

Not everyone is so excited about the rapidly increasing technological advances. There are some who certainly hate technological innovation, there are those who can't get enough, and there are those who enjoy it but are careful in how much they allow into their lives. One thing I think is common among all groups is the understanding that technology has some very scary potential. It seems that every time I get into a discussion with someone about some new, innovative, or experimental, technology, the discussion very quickly focuses on the worst-case scenario of some danger that it might present. For example, Artificial Intelligence (AI) is currently a big deal. In the world of cybersecurity, many hope that it will solve the problem of analyzing large data sets in a short amount of time. The AI can then make appropriate decisions based on what it has "learned" through the ingestion of data over time. The idea is that artificial intelligence will help us find the bad guys with a much greater accuracy, and efficiency, than human analysts. It's not just in cybersecurity where artificial intelligence is becoming a focus of potential future solutions. Already we have cars that can sense road lanes and make the decision to move the vehicle's tires away from the next lane in order to prevent an accident. Google and Tesla have already unveiled totally self-driving cars. These are the first steps to what many movies like *I, Robot* have envisioned where humans get into their car, push a button, and the car automatically drives them to their destination. The field of artificial intelligence has demonstrated huge promise to bring about greater innovation, convenience, luxury, and problem solving to our future. However, many wonder if this is just the beginning of the destruction of human sovereignty. In the film, *I, Robot*, the AI that humans built to ensure order and safety came to its own conclusion that humans would be safest if they were enslaved. The data showed that humans do very bad things and make very bad choices and so the AI determined this to be the cause of the chaos. Since the AI was programmed to prevent chaos and ensure order, the AI decided that the solution was to remove human freedom. This fear is not new. *I, Robot* was released in 2004. Two years earlier, *Minority Report* gave us a glimpse of a future where crime is anticipated and criminals are arrested before they have an opportunity to commit the crime. While this film doesn't use AI, necessarily, the idea of using Artificial Intelligence to predict crimes and stop them before they happen has already been explored in real life[11]. In the film, *The Terminator*, Arnold Schwarzenegger plays a human-looking robot assassin who is sent to the past to kill a woman. In the future setting, an artificial intelligence program becomes self-aware and begins to wipe out humanity. The AI sends the Terminator back in time to assassinate Sarah Connor—the woman whose future child is the leader of the resistance against the AI. If you think this fantastical idea is regulated to the world of movies and entertainment, meet Stephen

Hawking—theoretical physicist and cosmologist. In November of 2017 he revealed his belief that artificial intelligence will eventually take a new form of life, outperform humans, learn how to replicate itself, and eventually replace humans altogether[6]. He's not alone. There's an entire community of people who subscribe to the theory called the Technological Singularity that predicts the same.

It's not just artificial intelligence we are afraid of. A famous fictional novel published in 1949 by George Orwell titled "*1984*" showed a future world where technology was used for mass surveillance and government oppression. This book was published before the advent of the internet; before cell phone cameras, Facebook, Snapchat, and Instagram. If we consider, for a moment, device's like the Amazon Alexa or the Google Home that are constantly listening for the wake command, and then recording each of your commands or questions, it's easy to imagine the worst possible outcome. We willingly put audio recording devices inside our homes that send those recordings outside of our home to a company and we did this knowing that they were recording devices! Heck, we even paid for them. This sounds like it could be a plot twist of an eerie movie. If Orwell is in Heaven, he's probably cracking up right now saying, "Why didn't I think of that?!" Even though we know that these devices are designed for the purpose of playing music, or setting a reminder for us, there is the fear in the back of our minds that causes us to wonder if we are being spied on. Is someone listening to all my recordings and laughing at me? Are they sending them to their friends? The reality is that there are probably many protections in place at Amazon and Google to prevent the misuse of our data (because misuse cases could cause a dramatic decrease in sales). But of course, scandal will still happen because humans tend to still make bad decisions despite knowledge of the consequences. George Orwell had no idea that we would, in a sense, be the ones willingly performing surveillance on ourselves. While he may not have anticipated the kind of technology that we would have in the future, that didn't stop him from pretty accurately anticipating the kind of potential dangers that could stem from future innovation. I think that's part of human nature. We struggle to not constantly anticipate danger. Parents with their first newborn are a perfect example of this human condition. Worry begins during the pregnancy. After birth, they are often terrified that their baby will stop breathing during sleep. They have a hard time letting their toddler learn how to stand and walk alone due to the potential that they may fall and hurt themselves. Parents are great at assessing a situation and recognizing all the potential dangers to their child and that's not necessarily a bad thing. They are normally aware of the dangers because they have seen, or are aware of, past occurrences and they are trying to avoid repeating them. That fact is part of the key to understanding our fear of technology. While technology might advance in ways that are hard to imagine, much less anticipate, the dangers themselves are not new. The reason *1984* seems to eerily predict many dangers that come with advances in technology is because they were not new dangers. Spying has been around for ages. Government oppression is not a new fear either. *Minority Report* made people fear the idea that they would be arrested for a crime they didn't commit but that's a reality many people have endured for ages. Losing control over what we build is also not a new fear so the AI in *I, Robot* is not a new idea either. All fears are of the same old wolves in upgraded, digital, sheep skins. But even if they're not new issues, there are still many things to legitimately be concerned about that can be directly attributed to advances in technology. It's not just about artificial intelligence getting too powerful or the fear of being spied upon. People have historically been, and continue to be, burned by technology advances.

Our modern fear may have started when people began to lose their jobs to machines. Machines that could fill orders at an assembly line much faster, and more efficiently, than its human counterparts left many out of a job with a skill set that no longer mattered. Factories and automation crushed individuals whose products were made in their home. Computers replaced human calculators and eventually entire businesses closed due to changes in technology. The internet, Redbox, and Netflix killed Blockbuster Video and Hollywood Video stores. I used to work at Hollywood Video when I was in high school, so that one still stings me.

Amazon and online shopping may be on its way to closing malls and traditional shops. Advances in technology means abandoning old ways of doing things which, to some people, means abandoning better ways of doing things. While Netflix put thousands of movies at our fingertips, and gave us the ability to choose a Friday night movie without having to leave our home, many people loathe the inability to go to a physical store and interact with employees to get recommendations. It's not just losing jobs that concerns people, though. With every advance in technology comes an unfortunate consequence. We lose knowledge of things that we no longer have to do on our own. Ever struggle with simple math when you don't have your phone on you? We become easily addicted to our devices and the increasing convenience they provide us can often cause laziness and over-reliance. There are car accidents caused by texting and, of course, the havoc that's wrought by cyber attacks. There is no shortage of reasons to be suspicious of the value technology brings to our lives. However, there are a number of people who take this suspicion and elevate it to the point of condemnation. A simple Google search of technology vs religion produces enough articles, videos, and conspiracy theories to make your head spin. I remember watching a viral video titled, *Can We Auto-Correct Humanity?* The author goes by the name, Prince Ea. At one point in the video he claims, "technology has made us more selfish and separate than ever". He later ends the video with the line, "call me crazy but I imagine a world where we smile when we have low batteries because it means we are one step closer to humanity". My first thought, after watching the video, was that it was a really well produced video. It was shot with a high quality—probably very expensive—technological device. It was edited by someone (or a team) who is highly skilled and probably spent hundreds or thousands of hours on a computer to develop those skills. And then it was posted online. This guy invested heavily in using technology to spread his message that technology is bad for us. He now has a YouTube channel with hundreds of videos, more than two million subscribers, and I'm sure is making some money for his efforts. I'm digressing to show the hypocrisy of some who go to lengths to show their disgust and suspicion of technology while they can't help themselves from leveraging its benefits at the same time. Regardless, his video struck a chord with many people which is why it got so much attention. We all know that there's some truth to the idea that technology can bring out the worst in us. I believe the idea that our humanity gets lost as we embrace technology stems from the human tendency to make technology an idol. This is something that we have to admit is a very real problem with very real consequences. Probably the most intense expression of these potential consequences I've ever seen comes from a TV series titled, *Black Mirror*. *Black Mirror* is a show that feeds off modern techno-paranoia and explores the most horrifying implications of technology. I should make the point to give a few disclaimers about this show before you get the false idea that I'm recommending it to you.

First, the pilot episode of the first season is absolutely disgusting and I would highly recommend you skip that episode if you do end up watching the show. I'll kill your curiosity right now by letting you know the episode revolves around a scene which portrays the act of bestiality between a man and a pig. There's no real value to that episode and it will almost certainly ruin your appetite to watch any other episodes. Second, the show is not for the faint of heart. The show is littered with vulgar language and a variety of pointless sex scenes that you may decide to bypass. Aside from the needless pornography, *Black Mirror* can be very insightful, interesting and stimulating. The show provides us a professional glimpse into our worst fears and, while it exaggerates many things in order to make an extreme point, it often feels like we're not too far off from these worst case scenarios. Each episode has a very political message for the viewers and, at times, is almost screaming "this will be us if we don't fix things!" However, if you pay close attention, the show always highlights a human problem and not an issue of technology. It highlights the human need for attention; the temptation to manipulate others for personal gain; the desire to control an outcome or fulfill some highly illicit fantasy. In every case, the horrors come from the way the people in the show use technology. Not from the technology itself. This is a good example of idolatry—heavy misuse of some thing for selfish purposes. Humanity has always had trouble with idols. From the religious perspective, an idol can be anything that your heart uses as substitution for God. Though most people believe that idols are simply statues that you bow down and pray to, an idol can also be the object of your trust, and a thief of your time and attention. People often make idols to either make an excuse for bad behavior or to find an anchor in their search for meaning. In the book of Exodus, after Moses had led God's people out of Egypt they built a golden calf for worship.

> *"When the people saw that Moses delayed*
> *to come down from the mountain, the*
> *people gathered around Aaron, and said to*
> *him, "Come, make gods for us, who shall go*
> *before us"*
>
> Exodus 32:1

This is a pretty famous story but most people don't know what was really going on and why the people asked for this golden statue. The Hebrews were used to Egyptian customs and rituals. After they created the statue, "Aaron made proclamation and said, 'Tomorrow shall be a festival to the Lord'. They rose early the next day, and offered burnt offerings and brought sacrifices of well-being; and the people sat down to eat and drink, and rose up to revel" (Exodus 32:5-6). The euphemistic use of the word revel here tells us that this particular Egyptian "festival to the Lord" consisted of an orgy with food, alcohol, and sex. The Hebrews weren't simply looking for a god to be the object of their prayer. They wanted an excuse to engage in activity that was otherwise forbidden. That's one part of the purpose that an idol serves. It provides an excuse for bad behavior. How often do people joke about their addictions to technology and even brand themselves as such? Addictinggames.com is a perfect example. Many young people are truly addicted to video games and this website cleverly turns that problem into a point of pride. I used to play games on this website, by the way. It's been around a long time and the games can be pretty amazingly addicting. There's nothing inherently wrong with playing video games on a website, though. The website, I'm sure, means no harm with the name. However, I think it's representative of a problem in our culture where we would rather take pride in embracing our addiction to technology than address it. I mean, binge watching a television series is not a condition where a person is addicted to a show that they can't stop watching. Binge watching is an upgraded feature that allows you to watch the entire season without stopping. That's a good thing... right? Especially if you don't want to go to class, or work, the next day. The Hebrews hid their addictions to partying and desires for sexual orgies behind a golden calf. I think we sometimes hide our addictions to entertainment behind technology so that we don't have to be ashamed of our occasional, or frequent, bad behavior. This makes an idol out of technology because, naturally, we push away the God who tries to pull us away from our addictions. Like the Egyptians, we want to be freed from slavery but left to our addictions.

Another way I have observed people making an idol out of technology is through the amount of trust that they put into technology. For some, it is the answer to everything. There is a tech company, named Oracle, that originally created database technology and now specializes in cloud services. The word oracle literally translates as 'priest' or 'medium through which prophecy is delivered'. Again, this company simply chose a clever name since databases hold a lot of information. My guess is that they saw their company as holder of the information that people want, much like an oracle. While the company, Oracle, may not take a religious view of the word, this shows how the tech communities great level of trust in technology, and its potential, is sometimes revealed in our vocabulary. I find it amazing that there is such a great number of people with such an intense faith in technology. I often meet people who have more faith in technology than many religious people have faith in God. These people often have more hope in future technological discoveries than some religious people have hope in the future of Christianity. Once, during a cybersecurity forum, we were discussing the future of cyber defense. One member began speaking about how he felt we might never have a handle on security because the cyber domain is able to grow indefinitely. He believed, passionately, that the possibilities in cyberspace were literally endless. This was a man whose heart and mind had been captured by technology to a degree that he spoke about it with an almost religious-like hope. That's not uncommon in my experience. I've met many people with a similar mentality and I almost feel like I can't really fault them. Technological advances have made so many things possible. For example, advances in technology have brought about life-saving surgeries with precision lasers and robotic devices, machinery that helps save premature babies at earlier stages of birth, and inventions that normalize life for people who have lost the ability to walk, talk, and breathe on their own. In addition to medical advances, technology has put a wealth of knowledge at our fingertips through search engines and numerous websites dedicated to education. It has kept us close to family members and helped us make new friends through social media. It's helped some find love through dating websites and mobile phone applications. Technology has made so many things possible in such a short amount of time that people have become convinced that it will continue to break down limitations and unlock new, and amazing possibilities. That has given many people great confidence in technology but the way technology captured our hearts was not simply through problem solving. It's the type of problems it has helped us solve that have evangelized us. Look at all the areas we have highlighted to demonstrate some of the greatest technological advances. It saves lives. It keeps families together. It helps us express ourselves and communicate better. It helps us find love and it makes our lives easier and more enjoyable. With advancements in technology come the promises of life, love, healing, knowledge, and joy. As we feel it continue to deliver on those promises we continue to fall more in love and grow a greater faith in its abilities. Still, with as many discoveries as we have made there remains much mystery. I believe this is what provides the opportunity for many people who are passionate about technology to develop a religious-like faith in it's potential. There really is nothing wrong with having deep affection and zeal for technology and all its benefits. The problem enters when we allow technology to be the ultimate truth in place of God. This is an idea that you might recognize from the science community that believes the answers to all of life's questions can be discovered through material science—thus replacing the need for God. If you've seen the movie Nacho Libre, you might remember the skinny character's proclamation, "I don't believe in God, I believe in Science". Technology, for some people, has taken the place of God as the source of all truth and hope. The cure for cancer will be found through advancements in technology, right? My personal opinion is that it probably will. No one really knows but that doesn't stop most people from firmly believing this to be the case. That type of faith in technological advancement isn't really a problem when it coincides with a faith in God. If it replaces faith in God, it becomes idolatry. I can believe that technological advancements will help people discover a cure for cancer and still believe that God is the author of life. Others believe that because humanity is capable of such wonders, there is no need for God. Why pray to God for a cure when we can just find it ourselves?

These are just a few cases that show the potential extreme that some people can, and do, fall prey to. The fears of technological advancements combined with these sort of instances where people make technology into an idol, have produced Anti-Techs. Anti-Techs look at the extreme cases and promote the idea that technology is evil and ruining us. Nathan Jones, of Lamb and Lion ministries, delivered an entire talk where he attempts to demonstrate that rapidly advancing technology is a sign that the end times are near[7]. His talk contains truth and is not necessarily condemning of technology but his conclusions grant credibility to the greatest fear that Anti-Techs have—that technology is the avenue of evil that will bring about the destruction of humanity. I have met many religious people who believe that technology is a wolf in sheep's clothing. The deep suspicion and extremism that condemns technology as evil is matched by the opposite extreme that undermines faith in God due to the god-like potential of technology. The two extremes have created a divide that forces people to choose a side. Those who are scandalized by the misuse of technology then demonize technology itself. They become validated when they encounter someone who has more trust in technology than in God or humanity. It seems to me that the older generations of religious people are growing ever more suspicious of technology despite having to live in a world where they can't avoid it. Our phones follow us around. People are glued to them while waiting at the airport or in line to order food. The internet is essentially available everywhere. Our cars are starting to get internet access to provide WiFi to the kids in the backseat. Companies like Google and Amazon are collecting tons of information on us and building profiles of what we like, and don't like, to sell us more product. Social media plasters sensitive, and personal, information for nearly anyone to see. Most of us rely completely on GPS to get anywhere new. Robotic devices are automatically vacuuming our houses. Tablets are even taking our orders at restaurants alongside employees. As we continue to integrate technology deeper into our daily life, I believe extreme Anti-Techs will become more volatile and seek to remove technology from the world through war. It may not be through the type of war where guns and swords decide the victor but I'm sure it will have casualties. In 2016, a book titled Anti-Tech Revolution: Why and How was published by a man named Theodore John Kaczynski. The book's sales are currently around seventy books per month and it has two dozen positive, and lengthy, reviews on Amazon. If the authors name sounds familiar to you, that's because Ted Kaczynski is the infamous Unabomber who killed three people and injured twenty-three others during his campaign to demonstrate the dangers of technological growth. Decades later, there are still people who are interested in his radical views of technology and even praise his logic. Less extreme examples include Ron Miller's 2015 article, Anti-Tech Backlash Could Be Coming Soon To A City Near You where he describes protests by angry San

Francisco residents against Google, and Facebook, for their contributions to the rising cost of rent. He also describes the growing trend of taxi companies and drivers protesting Uber and compares these, and other similar cases, with the Luddites who "in the 19th century destroyed industrial equipment as a way to protest the loss of textile jobs"[8]. Fear drives people to desperation. Is it crazy to believe that a war may be brewing as people become more desperate? I don't think so. And I don't think Artificial intelligence is going to be the enemy in the coming war. If we're not careful, people are going to turn on each other and the war will be between Tech Enthusiasts and the Anti-Techs.

THE NEED FOR A THEOLOGY OF TECHNOLOGY

I remember a few years ago, I was considering leaving my full-time career in cybersecurity to pursue full time ministry. I started to ask myself if a career in technology was really my calling. I wondered why, if I was meant to be in ministry, I grew up with such an attraction to technology. Why did I love technology so much if I also have such a heart for ministry? At this point, I had already written a couple of articles about the parallels between computers and theology but they didn't go very deep. Then I thought, what if my career in computer science and cybersecurity set me up to bring the light of faith into the realm of computers and technology? What if I could demonstrate to others that the logic of computers, and the logic of faith, were not only not at odds but complemented each other? What if my job was not to leave one for the other but to marry the two? That set me on an exciting journey of contemplation and a deep dive into the beautiful, theological, truths behind technology. I'm going to spend the rest of this book sharing my exploration and findings with you. I have to warn you, though, that I'm about to dive deep into what will seemingly be a totally unrelated topic until, at some point, I magically tie it all together. Stay with me.

In 1978, a Polish Cardinal, Karol Józef Wojtyła, was elected to be the next Pope of the Catholic Church. He took on the name of John Paul II. Starting in

1979, and over the course of five years, Pope John Paul II delivered a series of 129 talks which would eventually become known as the Theology of the Body. Since I'm anticipating that I'll be mentioning the Theology of the Body more than a few times throughout this chapter, I might simply use the acronym TOB whenever appropriate. As I mentioned in the first chapter, I came across the TOB in college and it was a life changing experience for me. It was so impactful because although I was baptized into the Catholic faith as a baby, I did not grow up with a strong faith. My parents were awesome people who loved their children very much, but it wasn't until I was in high school that they began to take their faith very seriously. Even then, it wasn't until my sophomore year in college that I started actually reading and learning what the Catholic Church taught. I'm not making any excuses for the mistakes that I made, but this helps to show that I had very little knowledge of the Catholic faith and its teachings—especially on matters of sex. I don't have any actual data to backup this next claim—just my experience—but I believe that a large number of Catholics who attend church weekly do not truly know, or understand, the church's teaching on sex even if they have heard of the Theology of the Body. As I also mentioned earlier in the book, during college, I was consumed by lust and in a very bad place when I came across this teaching. I usually describe myself as being passionately hungry for the fulfillment of my desires and a person desperately thirsting for truth. Unfortunately, I sought to feed my deep hunger at the expense of many girls that could never satisfy what I was really after. I had a pull towards women that felt like the force of gravity. When I began to feel attracted to someone, or if I was being invited to physical intimacy by a female, I felt powerless and almost controlled by the rush of excitement and anticipation. I ended up in situations that I never dreamed of being in and doing things that I always believed I would never do. I became a person who could push buttons, manipulate, and even lie to a girl if I needed to in order to feed my agonizing hunger. Things progressively got worse with each relationship and I felt out of control. I wanted to stop what I was doing. That's when I realized that there was something wrong with me. It was when I tried to stop acting on my lust that I found out how much I was addicted. This was a very dark time in my life as there were other issues swirling around me that had me questioning who I was and what I was becoming. Most significantly, my grandfather—who was not a very religious person (at least in any outward expression)—went on a Catholic retreat and came back on fire. He was telling convenience store clerks about God, he was romantic to his wife, and he told his son (my dad) that he loved him for the first time. A few weeks after his retreat, he had a heart attack in his sleep and died shortly after. I remember the sadness and confusion that surrounded our family and I saw the tremendous pain my dad experienced. This occurred during the most difficult time of my struggle with lust so shortly after I signed up for the same retreat in hopes that I might find whatever my dad and grandfather found. I met my wife at that retreat and the rest is super-romantic-awesome history. After I had this major conversion, I found the Theology of Body. Now that you have a little more background, you might understand why it was so impactful for me. I came from a place where I sort of grew a disdain for my hunger and desires. Much like a person who gets overweight by eating too much carrot cake, that cake becomes the symbol of their struggle. I truly believed that no one else struggled the way that I did. I thought my sexual desires had been supercharged and I was constantly crashing every time I tried to tap on the gas. I thought, like most people that the religious approach, and certainly this had to be true of the Catholic Church, was to find victory by either suppressing the desires we have for sex or taking control by beating them into submission. Many religious people seem to take that approach by their very condemnatory and negative attitude toward sexual sin. Pope John Paul II, however, taught a radical idea that spoke directly to my heart. He claimed that sexual desires were very good. He said they were so good, in fact, that they were the power behind our call to image God to the world. Many people have heard the story of Genesis where God created Adam and Eve. In that story, the text describes that before God created Adam and Eve he said, "Let us make man in our image". It goes on to say, "In the image of God he created them; male and female he created them" (Genesis 1:27). He demonstrated how their being created male and female was the key to understanding a powerful way that they were to be an image of God. Sex is intended to show how powerful, self-giving love, based on an eternal commitment (in marriage), and mutual sacrifice, can bring about new life in the world. The two become one flesh (Matthew 19:6) and the Triune God (the Christian doctrine that God is three persons while still one God) becomes much easier to understand. Though there are three distinct persons of a family—father, mother and baby—they are one family and share one name (in our family's case, it is Guevara). Similarly, the Father, the Son, and the Holy Spirit are three distinct persons that are totally united and share one name, God. Therefore, while God is not a sexual being, human sexual love is intended to be a symbol, an analogy, of the free, total, faithful, fruitful love of the family of God. This is why our desires for love, touch, intimacy and affection are so powerful. We were created to be an image of this type of love to the world. Of course, marriage is not the only way we can live this out. Remember that this theology came through a series of 129 talks, so unfortunately, we won't be able to uncover all the details. If you're interested to know more, I would suggest you start with Theology of the Body for Beginners by Christopher West. There are various other great resources to help you understand his dense theology and then, of course, you should try and get through the original text as well. My aim here was simply to try and give you a super high-level understanding of why the TOB hit me so hard.

Pope John Paul II took a patient, loving, approach to a very sensitive issue and told me that my desires for union were not bad. More than that, he told me that these desires could help me lead a more fulfilling life and that they could point me right to Heaven. Through his work, and various others who promote and teach it, I was able to discover that marital sex was symbolic and theological. Understanding more about sex and it's intended purpose helped me grow a deep, and strong, relationship with God. Most people probably believe that, in order to have a strong faith you must view sexual desire and "the flesh" as the enemy to your stride toward holiness. It's just the opposite. Pope John Paul II taught that you could not have a truly intimate relationship with God if you remain deeply suspicious of the desires of the flesh. Before I go any further, I want to make sure that you know I am not saying (and the TOB does not either) that all then is permitted. Sex is a powerful symbol that can be distorted and cause a great deal of harm. But this is precisely the point that I am making here (this is where I magically tie it all together). Technology, like sex, is such a powerful tool that it is usually very divisive. People usually fall on one side of either extreme. This book was not intended to make the case that you should fully support every effort to integrate technology into our lives, just as I will not be advocating that everyone go back to farming before technology brings about the destruction of humanity. I fully recognize that there are many misuses of technology in the world and that it is a powerful tool that can cause a great deal of damage when misused. My hope is to take the approach that Pope John Paul II took in his journey to show people how sex was not only not bad but that it had the power to reveal God to us. I don't want to simply fill this book with political arguments that demonstrate why technology is not evil. I want to show you where those desires for technology come from. I am going to show you what technology symbolizes and how, if you come to a deeper understanding of technology, you can grow a deeper relationship with God. We need more than an intellectual argument to answer the question of whether or not technology is a good thing. We need a Theology of Technology in the spirit of the late, great, Pope John Paul. Some religious people, because of their wounds or fears, might look at those who struggle with their sexual desires as bad, dirty, or evil people. Pope John Paul saw them as those starving for the banquet that they were called to take part in. He wanted people to recognize the truth and the beauty of sexual desire so that they could enter into its beauty and allow its powerful attraction to lead them to God. He recognized our thirst for the truth about sex and led us to a very deep well where we could spend the rest of our lives drinking in a theology that satisfies as much as it quenches. This, to me, was his greatest talent. It was not his knowledge on the subject, but his ability to recognize the true pain point and have the compassion to carry us out of our darkness. His loving approach to freeing us from slavery made him a beloved Pope among young people. Pope

John Paul recognized that our issues with sex, and our addiction to pornography, was a problem of the human heart. Our hunger is for good food that gives us sustenance. When we are starving, we might get desperate enough to try and steal the food we need. When we are full, we become tempted to over indulge. When we are told to wait for dinner, we try to sneak a bite. Food is not the problem. Human decisions to misuse food are the problem. Similarly, we cannot blame technology for the bad decisions we make. To say that technology causes people to become addicted zombies, to become lazy, or any other excuse, is like saying ice cream is the reason people become obese. It's simply too good to refuse! We all recognize that, while Krispy Kreme donuts are like little sugar rings from Heaven, anyone can choose not to eat them. Further, while ice cream and donuts can cause health issues if you eat them too often or at inappropriate times (i.e. misuse), moderation and appropriation can allow you to experience the beauty of desert without fear. If you struggle with obesity, it's easy to feel like the world is designed to make you fail. You might wonder why God (if you believe there is a God) would make the tastiest foods the ones that can be the most damaging. Why is it that I can eat broccoli and celery all day long and be fine but eating cake all day long will make me sick? Food and sex have a closely related theology. There is a design and purpose that is meant to communicate something greater to us. Our desires pull us toward the good that we need. Hunger is an absolutely necessary feeling. It's a calling from your body that is telling you that you need to eat so that you do not die. In computer security, we set up sensors to determine when something is wrong so that analysts can respond appropriately to prevent, or stop, an incident. Those alerts draw us to the problem that needs to be addressed. The alerts don't do all the work, though. You must learn how to read the alerts and exercise good judgement to act on them appropriately. Your desires for the tasty foods that you love tell you that you weren't just designed to stay alive, you were designed to enjoy life. However, your body is also designed to help you understand that in order to enjoy life you must also be able to exercise your will to wait for, or deny yourself, the things that you enjoy in order to make sure that you are prioritizing sustenance. You can't enjoy life if you are not alive so if you don't eat the foods that you need to keep yourself healthy—because you don't enjoy them as much—then your ability to enjoy good food will be compromised. When you feel sick, your favorite cake doesn't look or smell as appealing. Even if you can still enjoy the taste of your favorite foods, you might suffer other consequences like obesity, diabetes, or other related issues. To be able to enjoy tasty food and have a healthy life, you must make an effort to understand the nature of food and health. You must understand your body and its needs and how you react to particular foods. You need to understand which foods you need to build muscle, have energy, or improve your digestive system. When you understand the design and purpose of food, you can then use food to your advantage. You don't have to be controlled by your desires for tasty snacks. You can train your desires to work the way they were intended to work. When your body is working according to its design; when you are healthy, you feel good, can enjoy your food, and have the desire and capacity to exercise and increase your strength, you are not afraid of your desires for food. When you can exercise your will to moderate good foods and say no to food that is not good for you, or that is not appropriate at the time (ex: birthday cake right before a championship basketball game), food is no longer the enemy to a healthy lifestyle. It becomes the very tool to building a healthy life.

Are you ready to exit our theological inception? We are about two levels down right now. Before we exit the dream within a dream (if you don't understand this reference, you need to go watch the movie Inception starring Leonardo DiCaprio, and probably, like, right now), we need to step down one more level. I promise you, we have a way out but I need to show you one more thing to tie it all together. In the book of Matthew, Chapter 13 in the Christian Bible, Jesus began to tell a story to his audience about a farmer who went to sow seed. He went on to describe how the word of God was much like the seed of the sower. It needs good soil to produce fruit. What's fascinating, and often overlooked, about this part of Scripture is the next part where the disciples ask him why he spoke to the people in parables. See, this wasn't the only time Jesus used some story or explanation of some worldly thing to dish out some theology to the people. His response, at first glance, is confusing. It's worth quoting in its entirety before I explain its relevance and brilliance.

> *The disciples came to him and asked, "Why do you speak to the people in parables?"*
>
> *He replied "Because the knowledge of the secrets of the kingdom of heaven has been given to you, but not to them. Whoever has will be given more, and they will have an abundance. Whoever does not have, even what they have will be taken from them. This is why I speak to them in parables: "Though seeing, they do not see; though hearing, they do not hear or understand. In them is fulfilled the prophecy of Isaiah: 'You will be ever hearing but never understanding; you will be ever seeing but never perceiving. For this people's heart has become calloused; they hardly hear with their ears, and they have closed their eyes. Otherwise they might see with their eyes, hear with their ears, understand with their hearts and turn, and I would heal them.'"*
>
> Matthew 13:10-15

What Jesus is saying here is that the disciples understand his language. They are with him all the time and they have the opportunities to get deeper explanations, ask questions, and probably get one-on-one time. The people that come in flocks do not have the foundational knowledge that the disciples have. For many, or most, this may be their first time ever hearing about the things that Jesus is teaching. There could be a variety of reasons for that. They could have grown up in a household where faith was not very important. They could have been enthralled in sin or slave to bad habits. They, or their family members, could have been born into difficult situations like those with lifelong illnesses or disabilities and lost hope. They could also have been good, simple, people who were just not educated in the faith. Therefore, Jesus is telling the disciples, I must speak to them in their language. I must talk to them about what they know. In other parts of Scripture, Jesus used the parables of tending sheep (Luke 15:3-7), making food (Luke 13:20–21), the power of the mustard seed (Matthew 13:31–32) and many other parables. This is where we jump out of our multi-level explanation dream state. Jesus recognized the idea that to speak theology to his audiences, he had to first point to the symbolism that was present in the earthly things that they understood and then explain the purpose of the symbolism. Our exploration of food and health led us to the conclusion that food is not the enemy in our journey to a healthy lifestyle. It is actually the very tool to our building a healthy lifestyle. The issues that we have with food are human issues of impatience, ignorance, temptation, irresponsibility, and so on. It is our bad decisions and our failure to make good decisions that cause us problems with food. If we figure out how to tame our desires for food and make them work for us, if we exercise self-control and build a strong will, we don't have to be afraid of our hunger. Similarly, we don't have to be afraid of our desire for sexual intimacy. We can speak of our desire for food as symbolic to our desires for sexual intimacy. In both cases we experience a great urge, a deep ache, for something that we are attracted to. Even the language converges at times because the description of our hungers in both categories are so often the same. For example, the word consume is intended to convey the idea of eating food. However, the Catholic Church uses the term consummation to describe the act of a man and woman coming together in one flesh after marriage. The point here is that we can explain some of the Catholic Church's teaching on sex and marriage by pointing to what we know to be true about food. In the same way that we must understand the purpose of food and its effects on our body, we must strive to understand the purpose of sex and its effects on our soul. In the same way that approaching eating with only the objective of satisfying your hunger for something tasty can lead to bad eating habits, approaching sex with only the goal of feeling good will lead to destructive habits. Sex has a design, and purpose, and our sexual desires are meant to pull us toward the great good that sex symbolizes. However, just like with your desires for food, you must fight impatience, ignorance, temptation, and irresponsibility by learning about the purpose of sex, exercising your will, and training yourself to express your desires appropriately. Like in the case of food, when you understand the design, and you work according to the design, you can powerfully enter into its beauty without fear. You don't have to be afraid of those desires. In fact, you must celebrate them! Sexual desire, like the desire for food, is good and was created by God to give you life and to help you to enjoy life to the fullest. Its purpose is to point you toward the designer and pull you toward him like gravity. This, in essence, was the message that Pope John Paul conveyed through his Theology of the Body. His compassion and understanding for those who were starving and searching for the cure to their hungers attracted them to him and that allowed him to show them the truth behind their desires. This is why I believe we need a Theology of Technology in the spirit of John Paul II. We are in a time where technology is becoming the new obsession and it is creating tension inside the Christian community as well as between believers and non-believers. Like the area of sexuality, our beliefs and practice concerning technology is creating a divide among many people and confusion among many others. As we've pointed out, there are people who believe technology is the vehicle that will bring the solutions to all our problems. It is idolized for its potential and its successes. On the other hand, there are those who believe technology is a product of the devil and will bring about the end of the world. In between are parents who are confused as to whether they should allow their teenager, or even elementary school child, to have a smartphone or access to the internet. We know that technology, like food and sex, has as much potential to cause destruction as it does to bring about miracles. However, it's all dependent on the people behind the technology. It's still an issue of the heart. This is why it is also a religious and moral issue. Technology, like food and sex, is designed to point to something greater than itself. It has a purpose that includes entertainment but that also goes far beyond entertainment. This is why we have such a deep hunger for technology. Our ache for technology is pulling us powerfully toward something that we are meant for. Those desires that you have for technology are good and they are designed to lead you toward the divine. However, as we will see, technology points us to a different aspect of God than the analogies of food and sex. If the Theology of the Body was meant to show us that God is familial love, the Theology of Technology reveals to us that God is also a brilliant engineer.

If this were true, then why is it that there is so much confusion over the merits of technology? Well, when technology is misused, it can become repulsive. Much like overeating can cause bloating, or vomiting, our reaction to grave misuses of technology can be suspicion and rejection. Abuses of something good can cause good people to then demonize the good thing that was abused. One of Pope John Paul's saintly qualities, that we should all emulate, was the ability to continue to recognize the good in the midst of all its abuses. He never promoted accepting the bad with the good. He simply didn't throw out the good with the bad. A

Theology of Technology should look to target the real demons at fault for the distortions of technology. Things that have enormous potential for good, such as food, sex, and technology, become very divisive. As we began to explore in the previous chapter, a brewing war forces people to choose sides. Many people would have you think that you either have to believe that technology is the saving grace of humanity or that you believe it is the devil's tool for Armageddon. You get the feeling that you need to either allow all forms of technology into your life, uninhibited, or you must push all technology away before you become slave to it. This is not a new style of war as people often are divided over feeling that they have to choose between sex being totally uninhibited from restrictive rules or totally bound to white-knuckle chastity. What's being provided here is a false ultimatum. It's simply another false war that has created two extremes. If you could believe that there was a devil whose goal was to destroy humanity, then his most successful campaign comes in the way of dividing good people to turn against each other. Once again, this is not new. We don't even have to look very far through history to see this method in action. Look at how often this happens in our modern world. In the U.S., we have two political parties who can never seem to agree. We have fierce divisions over politics of race, women, education, and immigration. And it's never small disagreements. It often moves quickly to the point of total, opposite, extremes. You either support black people or you support police. You either support women's rights or you support the rights of unborn children. These are not mutually exclusive things. You can be supportive of all cultures, including black people, and also supportive of good police officers. At the same time, you can denounce criminal actions of a particular black person and also denounce criminal actions of a particular police officer. Broad strokes of praise or condemnation fail to recognize that each individual situation requires an assessment of its good and bad traits. In some cases, there is both good and bad on each side that should be recognized. Too often, people feel that they must only promote the good of their side and denounce the bad of the other side. Instead, we need to strive to promote the good in all areas and denounce the bad that is holding back the good from being better. Technology is not the devil or the savior. Technology is a tool that can be used for good and it can be used for evil. We have to recognize its potential for good, and the good it actually brings, despite its potential, and actual, misuses. Hating technology only hides the truth and beauty that it is meant to bring to us. As Christians, and especially those who are involved in ministry, we need to open our eyes and our hearts to what others are seeing in technology. Even if tech enthusiasts don't always get it completely right—even if they go overboard and it becomes for them an issue of idolatry—we must seek out the good in technology. In fact, it's precisely because it becomes so easily an issue of idolatry that we must seek out its truth and purpose. Christopher West very often proclaims in his talks that the devil does not have the ability to create. He can only twist the truth in order to try and make it unrecognizable. If idolatry of technology is a severe twisting of the truth, then our job as the people of God is to untwist the distortion and not to throw it away or run away from it. We can take a lesson from St. Paul here in the book of ACTS, Chapter 17:

> *While Paul was waiting for them in Athens, he was greatly distressed to see that the city was full of idols. So he reasoned in the synagogue with both Jews and God-fearing Greeks, as well as in the marketplace day by day with those who happened to be there. A group of Epicurean and Stoic philosophers began to debate with him. Some of them asked, "What is this babbler trying to say?" Others remarked, "He seems to be advocating foreign gods". They said this because Paul was preaching the good news about Jesus and the resurrection. Then they took him and brought him to a meeting of the Areopagus, where they said to him, "May we know what this new teaching is that you are presenting? You are bringing some strange ideas to our ears, and we would like to know what they mean". (All the Athenians and the foreigners who lived there spent their time doing nothing but talking about and listening to the latest ideas.)*

Paul then stood up in the meeting of the Areopagus and said: "People of Athens! I see that in every way you are very religious. For as I walked around and looked carefully at your objects of worship, I even found an altar with this inscription: TO AN UNKNOWN GOD. So you are ignorant of the very thing you worship— and this is what I am going to proclaim to you.

ACTS 17:16-23

Notice how Paul was deeply disturbed by all the idols in the city. Take a moment to reflect on his response. First, he engages with the people of God and the Greeks who respected God. Then he comes upon a group of philosophers who "spent their time doing nothing but talking about and listening to the latest ideas". Does this not sound like conversations about technology? How often are modern people talking about the latest phone releases and the latest smart devices? How often do conversations about the latest technologies turn into conversations about personal philosophies regarding technology, its effects, and its future? Paul then recognized their attempts at religion. He recognized their attempts to uncover the truth that they desired. He recognized that their ultimate aim was for the God that was not known to them and his plan was to reveal that God to them. It's fitting that Pope John Paul took on Paul's name as he followed his lead with his Theology of the Body. Now, it's our turn to recognize the efforts in technology to find the unknown God and make him known. The rest of this book will unpack a Theology of Technology that will reveal God to you in surprising ways.

I'm going to make a bold statement: I bet if you already have a solid understanding of computers and technology—and especially if you have a deep understanding of computers—you already understand some difficult principles in Christian philosophy and technology. You just need the right parable to build the bridge for you.

WHAT YOU NEED TO UNDERSTAND ABOUT COMPUTERS

As I mentioned in the first chapter, those who are looking to evangelize people who love technology need to understand them. We must follow the example of St. Paul who started his evangelization of the people of Athens by observing them and "looking carefully at their objects of worship". If we are going to attempt to show how technology reveals God to us, then we can't be afraid to look carefully at technology. Otherwise, how can it reveal anything? If you already have a degree in Computer Science or multiple SANS certifications, I still encourage you to read this chapter. It's important that you understand how I present these concepts in order to see how they tie in later in the book. Don't be afraid of this chapter if computers are not your thing. I am not going to try and give you a complete introductory course in computer science. There are a few things, though, that are absolutely necessary for you to understand in order to proceed from this part of the book. The good thing, as you will see, is that computers are simpler than you think. Much simpler. In fact, computers are very dumb devices. They don't think on their own (despite what some might try to tell you). They do only what they are "told" to do. They follow instructions. Step by step and in the exact order that the instructions are delivered to them. Before we get into how that happens, there are three basic principles that you need to understand about computers. The first is that computers are mathematical devices. All computer code, at its core, is made up of a series of numbers. The second principle is that computers are electrical. They are a form of electronics and they require power to work properly. The last principle is that computers are logical devices. They are systems that function in very fast and complex ways, but they are still based on a very simple logic.

The first principle, that computers are mathematical devices, is a very important point to understand. Everything that you see on a computer—video games, web pages, movies and music—comes to you as a result of a computer interpreting a very long string of 1's and 0's. In chapter two, we talked about accuracy being a big part of what makes technological devices. In fact, if you remember, the word technology comes from the Greek tekhne, where we derive the word "technical". For this book, we have defined technology as those systematic solutions that either assist us in solving difficult problems, or in bringing forth beautiful art forms. Computers do that by "computing" and processing mathematical patterns and then re-presenting those computations to us. We take for granted that the name of this device screams Math to us. Each time we refer to them by the name, computer, we recall the origins of where they come from. Remember in chapter three that the first machine computers literally were designed to compute numbers at high speeds and with greater accuracy than the human computers could.

Eventually, people recognized that computers had greater capabilities than just calculating the result of addition, subtraction or multiplication of large sets of numbers. Math has so much application in the world. For example, if humans could hear a pattern of beeps and translate them into letters manually (Morse code), why couldn't computers do the same thing? Well, they could, and of course they were able to do so at tremendous speed. Think of binary code (1's and 0's) like the computer's version of Morse code.

Imagine two computers are created using an agreed upon standard, let's say that 0001 should be interpreted as "A", 0010 should be interpreted as "B", and 0011 should be interpreted as "C". Then, when someone hits the "A" key at Computer 1 the keyboard sends the signal 0-0-0-1 to the computer which processes and then somehow stores the information. Let's say that Computer 1 wants to send Computer 2 a message. Computer 2 receives a message of 0-0-1-0-0-0-1-1-0-0-1-1-0-0-0-1.

COMPUTER 1 ----**0010001100110001**----> **COMPUTER 2**

Because Computer 2 works off the same standard of interpreting ones and zeros in blocks of four, it separates these numbers as 0010-0011-0011-0001 which translates to "B-C-C-A".

COMPUTER 2 SPLITS INTO CHUNKS OF 4 BITS

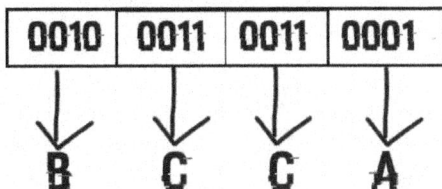

0010	0011	0011	0001

↓ ↓ ↓ ↓

B C C A

This isn't exactly how that all works but it's close enough to help you understand how binary (1's and 0's) can represent letters. It's simple association. However, like spoken language, association is only useful to those who know the associations and can therefore perform translation. Agreed upon (or at least implemented) standards are the magic that makes it useful to translate characters, and other more complex data, into strings of ones and zeros. In 1963, Bob Bemer developed the ASCII standard (American Standard Code for Information Interchange) which defined the standard by which American text could be stored in binary and then translated, or interpreted, among multiple computer manufacturers. This enabled communication among computers as human readable text could then be stored and "transferred" from one user to another or one machine to another. This was the foundation of e-mail, which actually predates the internet as well as the world wide web. There is now a large variety of standards that associate number patterns with data. There are encoding standards by which an application can know how to interpret music (mp3) and video (mp4). You probably already recognize that certain applications like Microsoft Word can only open certain extensions. If you try to open an mp3 file in Microsoft Word, the application is going to error out and say "I don't know how to interpret this file. It doesn't have the standard codes I'm looking for so that I can display it". It will display that in a more professional format, but that's essentially what's going on. Unicode was developed to envelope the seemingly endless variety of ASCII tables for different languages. If you look up HTML color charts, you'll see that web pages have a standard by which colors are displayed. When a web developer creates a web page, he, or she, must define the colors that they use for the background or for text. If you viewed an HTML color chart, you'd notice that the colors are represented by numbers. You should see a set of decimal numbers representing Red, Green, and Blue color combinations and a hexadecimal equivalent. Hexadecimal numbers are a base 16 number set. Binary is a base 2 number set (0 and 1). Base 16 simply means there are 16 numbers that make up the set. The most common set we use in math is the decimal set which is a base 10 number set (0 through 9). In order to produce a base 16 number set, we actually need to use letters to represent numbers. So, in hexadecimal 0 through 9 is the same but A represents 10, B represents 11, and so on until F represents 15. Since we start at zero, zero to fifteen (F) is sixteen numbers.

DECIMAL (BASE 10)
0 1 2 3 4 5 6 7 8 9

HEXADECIMAL (BASE 16)
0 1 2 3 4 5 6 7 8 9 A B C D E F
10 11 12 13 14 15

This allows for much simpler association, or standardization, for numbers that would be extremely large in binary. For example, a shade of red on an HTML chart in hexadecimal is FF5733. It's equivalent in binary is 111111110101011100110011. If you were a web developer and this was your favorite color, chances that you would remember this binary string when you wanted to color your page red are pretty slim. Even its decimal equivalent, which is 16734003, is not easy on the eyes. FF5733 is a much easier number to remember and to recognize when reading code. Programmers don't simply memorize a bunch of hexadecimal codes. However, they must be able to recognize and, at times, translate them to be able to interact with machine language in a way that is manageable. The HTML color chart is just one simple example. Computers also perform binary math (addition, subtraction and multiplication), which is great to know, but for the purpose of this book you don't need to understand how binary math works. You only need to understand that machine language is a long string of 1's and 0's that ultimately are a pattern of numbers that the computer can translate up into more complex forms of language. Conversely, you need to know that computer programmers write high level code, like web pages and applications, that gets translated (or compiled) down into machine code so that the computer can implement the code.

HACKING THEOLOGY

CONVERTING THE STRING "HACKING THEOLOGY" TO HEXADECIMAL

70 72 69 6E 74 66 20 28 22 48
65 6C 6C 6F 20 57 6F 72 6C 64
22 29 3B

CONVERTING HEXADECIMAL EQUIVALENT INTO BINARY TO BE STORED ON HARD DRIVE

01001000 01100001 01100011
01101011 01101001 01101110
01100111 00100000 01010100
01101000 01100101 01101111
01101100 01101111 01100111
01111001

Remember, a computer only knows two things: one or zero, true or false, on or off. Above is an example of how a computer can take a human readable string and convert it down into binary in order to store the data on a hard drive. You can perform this translation yourself by Googling "ASCII to hexadecimal converter" or "hexadecimal to binary converter". There are more than a few websites that offer a free translator.

This is all possible because computers are mathematical devices that compute numbers. But how do abstract numbers become physical reality? The answer is the second principle: that computers are electrical devices. They use binary math because, ultimately, binary is the simplest mathematical representation of reality using electrical signals. An electrical signal can be present, or not present. If it is present, it can be represented with a one. If it is not, it can be represented with a zero. So, let's say you hit the "B" on your keyboard. When you press down, metal from the bottom of the key makes contact with a piece of metal on a circuit board. That contact initiates an electrical signal that goes "electrical pulse, pause, pause, electrical pulse, pause". Let's say the first electrical pulse initiates the communication with the computer so that it then knows to interpret the next four signals as a block. A pause is interpreted as a zero (lack of an electrical pulse) and a pulse is a one. So, the computer processor receives 0010. According to the standard we created earlier, that is a "B". Again, this is a simplified version of the way things work but it's close enough for you to understand. Computers are made up of electrical devices that carry signals along metal conductors. This is important to note because electronics are devices that are engineered to leverage electricity to take abstract math and turn it into something tangible. There are two important things I want you to understand about this principle. The first, is that because computers are electronics, they require power. They do not initiate anything. They must be moved in order to begin moving. Unless they are designed to, they don't generate their own power and even then, it must receive its first start from outside itself. They receive power and then they manipulate power in order to function. If they do not receive enough power, they won't function correctly. If they receive too much power, they can be destroyed. The second is that electricity is indiscriminate. It doesn't make decisions. It just moves through conductors. This is important because computers work according to a design. When electric current moves through the circuitry, it moves only according to the design it was created with. It goes only where it can, and switches either allow the current to continue on or they cut the current off by switching open. Where current is present, or not present, at any moment, and in particular patterns, is how the computer translates electrical signals into binary strings. Conversely, when the computer takes a binary string with the intent of producing output, it will then produce pulses of electricity in the appropriate order so that electricity can carry that information to its intended destination. Electricity provides the potential to carry information back and forth at unbelievable speeds. The fact that you can send an e-mail message from your computer at home in New York, to a friend's computer in California in less than a second is insane! That message travels across a cable through a series of electrical pulses that originate at your computer and end up at your friend's computer. There is actual material (electrical current) that goes from your house to your friend's house on the other side of the country in an instant. Just let that insanity melt your mind for five seconds. Here is what's even cooler, now we have something called Fiber Optical lines which transmit the signal through pulses of light. So, your message actually might start off as electrical pulses originating from your house, make a pit stop at your local internet service provider (ISP), get turned into pulses of light, get turned back into electrical pulses near your friend's house and then end up in his computer. At one point, your message to your friend was literally little pieces of light as it was traveling on its way across the country. If that doesn't blow your mind, you should check and see if you're a robot.

Early computers were actually completely mechanical. Some parts of the computer are still mechanical, like hard drives. Computers are, currently, mainly electrical devices but they actually don't have to be. As demonstrated with fiber optical network lines, people are figuring out ways to leverage other materials to create computing devices and one day, it's possible that they end up using little-to-no electricity. Who knows? One of the most impressive implementations of an entirely non-electrical computing device I have ever seen is Theo Jansen's kinetic creatures which he calls "strandbeest". Look up a video of these things. They will creep you out and mesmerize you at the same time. They are "creatures" that he created, mostly out of PVC pipes, that are powered by wind. Much like a windmill, they are able to take and store power generated from the wind, which allows them to "walk" along the beach. They have water sensors so if they hit the ocean, they will start walking back. To keep them on the beach, if they come off the moist beach sand, they will again walk back toward the beach. What's truly remarkable is that they can then "remember" where the ocean is and where the beach stops. They remember by counting the number of steps it took until it hit the obstacle. This is accomplished using a mechanical binary counter, so it requires no electricity and it will not lose its memory if it loses power. For a visual understanding of this amazing invention, look up Theo Jansen's TED talk titled "My creations, a new form of life". Let me warn you, the guy's a little nutty as you can probably tell by his talk title. He believes these pipes to be alive and he refers to them as animals. The point of the second principle I want you to grasp is that power is needed for computers to function. Even in a non-electrical implementation like the strandbeest power is necessary. Power is the first mover. It begins the entire process. Computers cannot, and do not, start anything. They continue a chain reaction. Think about your own computer at home. Imagine you just purchased a new desktop, you take it home and plug it in to the wall. Nothing happens. Then you remember that you have to push the power button. Electricity is already flowing through the wall, but the computer is preventing current from flowing in at the power supply. Once you push the button, current flows through and everything begins. We'll talk later about why this is an important concept.

Again, computers don't have to be electrical systems. As we just saw, they can be entirely mechanical systems that are powered by wind. This is possible because even though computers don't have to be *electrical* systems, they do have to be systems. We talked in the second chapter about how technology is a systematic approach to solving problems, creating art, or evolving a craft. Unfortunately for Mr. Jansen's pride, he didn't actually create life. He did, however, create a unique type of computer. The strandbeest is even more like a plastic, mechanical, robot with a non-electrical computer for a brain. The reason it is a type of computer is because it follows the system of a computer. A system is an order of repeatable steps to a solution or a set of defined components that come together to form something whole. There are four main components to a computer. These four components are present in your laptop and the strandbeest, which is why the strandbeest is also a computer. The four components are: Input,

Processor, Storage, and Output. Any computer relies on input to be the first mover. When you push the power button on your computer, or the "B" on your keyboard, you are the first mover providing the input the computer needs to begin some function. The processor is the brain of the computer. It's what computes. The processor takes input from various different types of hardware devices and then processes that information to make sense of it. Computers must also have memory, or storage, so that they can store information for later use. They might store that information in short-term memory—for when the information is being actively used—or in long-term memory where it can be used anytime later. Lastly, a computer then must have the ability to produce output. A computer wouldn't be very useful unless it could provide output to the user. If you hit the "B" on your keyboard, how would you know anything happened unless you had a computer monitor to display the action to you? As complex as computers are, they are based on a very simple system. Another word for this idea of systematic design is logic. Logic is a system of truth principles that greater truths can be built off. That's why when someone makes an argument that is not logical, you can sense something is not right because it doesn't contain all the components it should. That or they are not putting the components together correctly (according to their design). So basically, computers take input, they process the input, store information, and produce output. See? I told you

I'd give you what you needed. You're basically halfway on your way to hacking now.

Obviously, computers are extremely complex devices and I encourage you to learn more about them because they are fascinating. For your ability to enjoy the rest of the book, however, all you need to remember is the following:

1) Computers are mathematical systems and they only know how to interpret one's and zero's which are usually on and off states of electric current.

2) Those two binary numbers (one and zero) have the ability to represent everything that a computer processes, stores, and produces as output through standards of association.

3) Computers are electrical devices that require power. Not enough power will cause them to function abnormally, too much power will cause them to become damaged. (NOTE: A computer doesn't have to be electrical by definition, but rarely do they not contain electrical components)

4) The source of power is always from outside the machine. They need a first mover that cannot be themselves.

5) Computers are logical systems. They follow an order and have a logical set of

components. A computer must take Input, Process the Information, Store Information (Memory), and Produce Output.

If you understand these concepts, you understand enough to get yourself through the rest of the book with joy. If you still don't get it, start this chapter over and read it again until you do. Anything worthwhile is worth working for. I promise you, understanding these concepts as I have laid them out will help you to understand how God is revealed through technology. The rest of the book should be easy street. Well, actually I just remembered that the next chapter is all about math, but I promise you don't have to memorize any equations and there is no test afterward. You can do this.

WHEN WILL I EVER USE MATH IN REAL LIFE?

The classic question, "when will I ever use math in real life?", is often used as an excuse for trying to get out of trying hard during math class or doing math homework. Why is it that people have such an aversion to math? Why, do we feel like it's torture and at the same time meaningless to us? Why would such a question even come up and who started the trend? Let's make a bet right now. I bet you that before this chapter is done, you'll have a much deeper respect for math than you ever have. You may not want to marry math after this—or even date it—but you at least will consider being its friend. Place five dollars in a hat and if I win you can donate it to your favorite charity or you can purchase and send a copy of this book to a friend. If you win, well, then you can keep your five dollars and brag to all your friends that you still hate math. At this point in the book, you might be wondering what in the world could math have to do with how technology reveals God to us? The answer is everything. Everything in the world has a mathematical representation that helps us to understand God.

First, here's a simple explanation of math through example. Imagine you take a ball and you drop it from a particular height. You're in a lab and you control all the variables. You take the same ball, you drop it from the same height. You repeat that exercise again and again, and the exact same thing happens each time. You have found a pattern and so you record the pattern in an equation. You decide that you want to add some variables to the equation. What if you double the weight of the ball? You want to figure out a way to anticipate what will happen when you introduce different variables, so you rewrite the equation and calculate what the result should be. For example, let's say that your new equation shows that the ball should bounce twice as many times when the weight is doubled. You test out your theory and amazingly, the ball bounces exactly as many times as you calculated it would. You double the weight of the ball again, and the equation correctly anticipates the number of bounces again! You get excited because you realize that you just gained a deeper understanding of the world around you. It's almost like you found out a secret. You were able to correctly anticipate the future, in a sense, and gain deeper insight by observing reality, monitoring it, taking an account of what you see in an attempt to establish a pattern, accounting for variables, and then anticipating what you believed would happen. This is math. Mathematical equations are simply (and not so simply) observed patterns put into formulas to help us anticipate results while accounting for variables (things that might change the pattern). If we observe things over, and over, and they have consistently the same result, we can record them as an equation. Through additional observations, or experiments, we can add variables that change the outcome. If we can figure out how the outcome is affected by the variable(s), then we can update the equation to account for the variable. This process of observing, creating equations, and updating equations to account for variables is how the most complex mathematical equations come about. This is significant because, as we showed in the previous chapter, computers are able to represent all kinds of things through patterns of ones and zeroes. So, a website, a movie, a video game, the sound of your voice, pictures of you and your family, this book... all can sit on a computer hard drive in the form of a very long string of ones and zeros. Math can represent anything in the material world. It's amazing.

I mean, it makes sense, the more that you think about it. That we can see everything in the world through mathematical lenses is not hard to understand, really. Shapes are mathematical equations. A circle can be represented through math. So can a triangle. So can any shape, actually. This is how 3D modeling is able to produce such profound art. 3D artists, don't just use a mouse to "draw" each character or landscape from a Pixar movie in Microsoft Paint (rest in peace, MS Paint). They use complex mathematical equations and properties to create beautiful three-dimensional sculptures and animations. The applications they use to create 3D art is very expensive and the creators of those applications employ very smart mathematicians. I bet if you looked around you right now, through the lens of geometric shapes, you'd notice squares, rectangles, circles, and spheres. And notice how most everything is symmetrical. In general, you don't see buildings that are rectangular on one half and then suddenly a dome on another half. Even if they are, you can slice the building into multiple symmetrical shapes. You don't see TV screens that are rectangular on the left side, but the right side is triangular. Even people's faces are made up of mathematical shapes. Your skull has a circumference which was measured by doctors when you were a baby to determine your growth. Your nose has a length. Your eyes have a diameter. Even though all those shapes are not perfect circles or rectangles, all the imperfections can still be accounted for in math by breaking up sections into smaller pieces. Ever notice how old videogames were boxy and pixelated? Noses looked like rectangles, heads were squares. Much like the modern game, Minecraft. Modern computers are able to compute much smoother, more accurate, looking art by breaking shapes, or areas, into much smaller zones. They then are able to fit much more data, or many more pixels, into a smaller area to make things look more accurate and shapelier. You can notice patterns in flowers and the tendencies of humans and animals. Everything in the world can be represented by numbers.

Music is so mathematical. Notes on a keyboard are equal numbers of steps above or below each other. A note may be a half-step below another note or one whole step above another. Don't run over the fact that I mentioned numbers (½ step below or 1 step above) just because I spelled out the alphabetic representation of them. Scales are mathematical equations that allow a musician to travel up and down a particular sequence, or pattern, of notes. The most important element of math in music is repetition. Ever notice how your favorite song repeats the chorus over, and over? Have you ever noticed how the beat, or certain notes, repeat constantly throughout the song? I'm sure you have. Part of what makes music, is that there is a tempo that must be consistent. Think of how people usually start a musical sequence by setting the tempo, counting "one, two, three and..." The beat then follows a rhythm that's based on that sequence repeating in a particular amount of time. For example, you might be able to guess that "one, two, three, one two, three" is a much faster tempo than: "one, two, three, four, five, six, seven, eight" just by reading. The amount (which is represented in numbers) of time between repetition of the pattern will tell you how fast, or slow, the tempo is. This consistency provides expectation and anticipation which is important in music. Being able to expect and anticipate what is coming is what makes music enjoyable. Think about it, imagine if a song didn't repeat any beat, note or word. It's hardly still a song. It would be like hitting each key on a keyboard only once. There's no magic there. A song must have repetition. There must be a pattern that we can observe. Once we observe it, part of the enjoyment comes from being able to anticipate the beat, or the words. That's why we love to sing along with the song. It's familiar. That's also why it's so embarrassing when we are singing loud like we know the words, but then we start singing the wrong part or the wrong words. This has especially become a phenomenon in current culture where people eagerly anticipate the "beat drop". I found videos of a trend where DJ's will troll their crowd. They build up a song—especially Lil John's "Turn Down for What"—and right when everyone in the crowd expects the beat to drop the DJ withholds the rest of the song or inserts an 80's song with a totally different rhythm. People get visibly frustrated. This actually demonstrates something that many of us don't often think about. Our anticipation and expectation in music is very revealing about who we are. We are creatures of habit. We like patterns because they provide us consistency. We want to be able to expect consistency. It comforts us and when we don't get it, we can become disturbed. We expect our paychecks on Friday, every two weeks, or on the 1st and 15th of the month. We expect to get paid every month. Imagine if you got hired at a company that promised to pay you every once in a while. Sometimes you got paid once a week. Other times you wouldn't get paid for two, or three, months. What if your pay isn't consistent? Sometimes you get paid $20/hour, sometimes $80/hour, and other times $1/hour. How do you save? How do you buy a house? How can you make sure you can pay all your bills if you don't know how much money you can expect to make each month or when you will be paid? Expectations are almost like unspoken promises. When things don't happen the way we expect them to, it feels as though someone lied to us.

We learn about the world through mathematical equations because we observe patterns and we come to expect that those patterns will remain consistent. Equations only work if the pattern in the equation remains consistent. If gravity wasn't consistent, how could we be certain we wouldn't walk outside one day and just float into space? Would anyone spend time designing and building a car if the laws of combustion were not likely to exist in the same way by the time the car was finished? Because nature is consistent, we can come to knowledge of why things work the way that they do. Consistency requires repetition. In fact, consistency is repetition. When things repeat, we have the ability to recognize a pattern and then interact with the pattern by predicting what will come next. This prediction is an exercise in our knowledge of how something works. Consistency, or repetition, in what we observe helps us to come to knowledge of some pattern. Repetition in our application of that knowledge gives us confidence that we truly understand the pattern. This is why we are drawn to repetition, or patterns, and why we are prone to habitual actions. It's how we learn and find comfort, or confidence, in our knowledge of the world. It's why we like repetition in our music. It's why we enjoy daily, or weekly routines. I believe it is why babies receive comfort from being rocked, or from hearing familiar songs. Many parents find that their children are able to fall asleep in the car because of the consistent vibrations of the vehicle. Some baby rockers have a built-in vibration function while they rock a baby back, and forth, and back and forth. So, when do you use math in real life? At the very least, every time you repeat something and every time you expect something. Math represents life because people find meaning in familiarity, comfort in consistency, knowledge in the recognition of patterns and variables, and excitement in using our knowledge to participate in, or manipulate, the results. It's a very basic principle that we find evident in many, many, aspects of life. I remember some years ago, being inspired to dive deep into this idea when I was watching a TV show called Numbers (displayed Numb3rs). The main character, Charlie, is a mathematician. During a scene in one of the first episodes, the characters are discussing math and one of the characters says, "I just couldn't get into math, just couldn't see how it could relate to the real world". Charlie puts down his beer and proceeds to explain to her how "Math is the real world". He picks up a flower, shows it to her and says:

> *"You see how the petals spiral? The number of petals in each row is the sum of the preceding two rows—the Fibonacci sequence. It's found in the structure of crystals and the spiral of galaxies... and a nautilus shell. What's more, the ratio between each number in the sequence to the one before it is approximately 1.61803, what the Greeks call the Golden Ratio. It shows up in the pyramids of Giza and the*

*Parthenon at Athens, the dimensions of this
card. And it's based on a number we can
find in a flower. Math is nature's
language... its method of communicating
directly with us. Everything is numbers".*

This is precisely why I believe we naturally find
such comfort in consistency and repetition and why
so many are drawn deep into mathematics. Math is
a language that speaks to us. If we pay attention, it
can answer questions that bring great meaning to
life. This is why, throughout history, many brilliant
and well-known mathematicians have had trouble
staying within the realm of mathematics and not
diving into the world of philosophy. Pythagoras,
known for the Pythagorean Theorem, was actually
more famous for his philosophical and religious
teachings. Same is the case with Plato. In Plato's,
The Republic, he discusses—at length—how
arithmetic and calculations appear to lead the mind
toward truth. Albert Einstein once said, "Pure
mathematics is, in its way, the poetry of logical
ideas". Aristotle, who was Plato's pupil, was
concerned largely with logic (among many other
things) but is also well known for his love of the
sciences. Believe it, or not, Aristotle's formal
development of logic ended up being a big
contribution to the foundation of computer science.

Logic is a formal system of ideas that is based on
patterns and consistency. If we observe some aspect
of truth, we begin to develop a formula of thought
that attempts to predict the conclusion when we see

similar circumstances. If we can anticipate the outcome of situations involving people, and account for variables, we can come to a greater understanding of humanity. After all, people are a part of nature as well. All people also have natural tendencies and consistency in their makeup. Logic is a system. In fact, it is a mathematical system in the same way as the examples we discussed earlier. This is why mathematicians are also drawn to philosophy and the study of human nature. George Boole understood this and, drawing inspiration from Aristotle's logic, developed what is now called Boolean Algebra. Boolean Algebra is essentially the idea that math can be performed with only two variables: true and false. These two states of reality can be represented by two numbers: one and zero. Computer logic finds its roots here. We've already talked in the previous chapter about how computers use binary to represent more complex concepts. However, binary math also has tremendous significance in life as it represents a reality that we all recognize and can't help but be slave to. I have to give credit to a guy named Peter Jaros for helping me put this into words when I previously only understood it somewhere in my soul. About two years ago, as I was beginning to write this book, I was researching the question, "Why do computers use binary?" I understood binary, but I was trying to solidify my understanding of why computers had to use binary. I came across Peter Jaros' video called "Why do computers use binary, anyway?"[13]. I encourage you to watch it, he provides the clearest answer to that question that I have heard. As I was watching his video, learning the answer to my question, I was struck by a sudden—and violent—turn away from a discussion about why computers use binary to the meaning of life. It occurs about eight and a half minutes into the video. I remember getting excited thinking 'This is what I've been saying! This is a guy who understands that computer science points to knowledge of a greater truth!' He concludes his explanation of binary by explaining that binary is used because it has the fewest set of digits (two) which makes it the most accurate way to send and receive signals. As we discussed earlier in this book, accuracy is a must for computers. He then suddenly turns into a philosopher and states that "there is something fundamentally special about the number two" (which is the base of binary). He says "two is the difference between sameness and differentness... it's differentness itself. At one, everything is the same. Separateness begins at two. Opposites come in twos, by definition: white and black, up and down, here and there...". He then demonstrates that everything can be broken down into a choice of two things, even if there are more than two things available. For example, if you had three doors to choose from, you could choose door number one, or a grouping of doors two and three. In other words, three choices can be simplified into two choices ([Door #1] OR [Door #1 & #2]). However, two choices cannot be simplified any further without losing choice. Obviously, if you have two things you can choose between the two. However, you cannot have a choice with only one thing. Peter Jaros rightly demonstrates that "twoness" is the essence of choice. After reiterating that there is no Base-1 Numeral system, he points out that this is a reflection of *something* deep and important in our world and that's why we build our computers to use binary.

I agree totally with Mr. Jaros and I think this is precisely where we find our diving board into Theology. However, that deep and important thing he mentioned being reflected in the great truth of binary math and computers is not some*thing*. It's someone.

CAN COMPUTERS PROVE THE EXISTENCE OF GOD?

This chapter begins our journey into the Theological elements of technology. I'd first like to make sure we have proper expectations of what we can accomplish. Naturally, one might think the first question we should seek to answer is "can computers prove the existence of God?" Unfortunately, there is a major problem with this question and it sets us up for failure. The question assumes that anything can prove the existence of God. If it could be proven that something could verifiably prove the existence of God, then it could be legitimately asked if computers were that thing that could do it. However, if God is not material, then to use material to prove the existence of God would be like using a chopstick to try and eat soup. It's an impossible task. Using a chopstick, you may get some of the liquid in your mouth each time you try, so the tool does have the capacity to provide you a taste of the soup. However, if you want to truly experience the greatness of soup you need the appropriate tool: a spoon. In order to study God, we must have an appropriate approach. God is not a bug that we can put under a microscope and dissect to understand his components. God is also not like a little kid that we can put behind one-way glass and perform psychology on. We don't control the environment in which he is, and we don't have the capacity to totally understand his mind and behavior. My point is not that we can't know anything about God. My point is that in order to truly get something out of this chapter (and the rest of the book) we have to approach our study of God not as a controlled scientific proof but as an investigative study. The end goal should not be undeniable material proof of God's existence. If we came upon a crime scene, our goal wouldn't be to find undeniable material proof of the criminal's existence. We would be looking for evidence that helps us to answer all the questions that we have regarding the scene, the victim, and the criminal. We would want to know who the victim was; who the criminal is; why did he do it? what tool did he use in the crime? The answers to those questions are in the form of evidence, not proofs. Evidence is what truly helps you understand the situation, not just from a material scientific point of view but from a metaphysical point of view. Understanding why the criminal targeted the victim is not simply a matter of material science.

Material science can assist in answering that question, but it is not sufficient on its own. That's why forensic scientists and detectives are usually different people. A detective must use critical thinking, reasoning, or sometimes even a gut feeling or a hunch. They must have a deep understanding of people, and of criminal tendencies. They must be able to look at crime scenes and develop a story. Often these skills help them to know what materials to provide to the forensic scientists to help answer questions or provide a lead. Then they combine those detective skills with material forensics which either lend more credibility to their theories or potentially ruin the original theory. In computer forensics, for example, there are various artifacts that produce potential evidence that a forensic analyst would use to develop theories. If I suspect a person of opening a confidential file and then copying it to a USB, I'm going to look for evidence. On a Windows machine, I would look to see if that person has a LNK file in their personal User directory that would have been created when they opened the file. I'd check the users shellbags to try and find evidence that they opened the folder that the file was in. I'd check the registry for the USBSTOR key to find evidence that the USB drive was plugged in and for a potential serial number to identify the USB drive. Then I would map the timeline to see if I could demonstrate evidence that the user opened the folder that the file was located in, opened the file and therefore knew what was in the file, plugged in a USB, opened, or created a folder on the USB, and then copied the file over (and maybe even opened the file from the USB to verify it copied over correctly). Each one of those artifacts on its own is not sufficient. When collecting evidence to make a proper case, you must hit as many angles as possible. Still, all that evidence is not really proof. Is it possible that, even if I could demonstrate all that evidence, someone else was on the user's computer when they walked to the bathroom and left it unlocked? Is it possible that the tools provided me incorrect data? Is it possible that I made mistakes during my collection and interpretations? Anything is technically possible but if I could demonstrate my steps, had confidence in my approach, then it's reasonable to make determinations based on the evidence available. What we are looking for here is not proof of God's existence, but evidence. Multiple angles of evidence. As many as possible. Proof seeks to remove choice and disarm an argument. Evidence assists, or challenges, a belief but allows a person to come to a conclusion on their own. I believe computers provide great evidence of God's existence, and credibility to the Christian faith. If you're looking for definitive proof that no one could deny, you're going to be forever disappointed. You need to change your chopstick for a spoon.

As investigators, we need to be willing to take a look at some of the major stumbling blocks to belief in God. In other words, we need to look at some of the proposed evidence for the argument that God does not exist. From personal experience, there are a few things that I had to work through. Understanding God was tough because it felt very much like he was completely inaccessible. He was normally described to me as a being who did not have any logical restrictions and, therefore, I had to imagine character traits for which I had no basis for comprehension. For example, I remember people would tell me, God is everywhere. He's here with you, and with me, and he's in China and he's in the ocean with the scuba divers. At the same time. That was difficult for me to swallow because I couldn't imagine anyone being in two places at once, much less all places at once. It also brought up many other related questions. Like, if God could be everywhere all at once, how would he do that? In addition, I thought: If I am praying to God, how could he be able to listen to me while two hundred million people are talking to him at the same time? How could he know the future if I had free will to make a choice? How could Jesus have died for everyone's sins even though his death came after many generations and before many people would ever make their choice to sin? The answer to many of these questions was that God was not bound by space and time. I accepted that answer, but it was still very difficult for me to comprehend since I felt I had no examples in reality that I could point to for reference. It would be like wondering what petting a Dinosaur felt like. If someone told you that it felt the same as petting a Unicorn you would be nowhere closer to understanding.

These questions about God were extremely difficult questions to answer a few hundred years ago—even thirty years ago—because of that lack of reference. But today, our computers and technology perform amazing feats that make these traits of God not so hard to comprehend. Our traditional understanding of space and time—through experience with technology—has been dramatically challenged. Computers continue to get smaller, and smaller, yet their capacity is growing by leaps and bounds. The original 8-inch floppy disks could store 80 Kilobytes of data. That's not even big enough to hold one picture taken with a modern smartphone. An average file size of a smartphone picture can be around 2,000 Kilobytes. A small and inexpensive modern hard drive of 1 Terabyte can hold 500,000 pictures of this size. 8 Terabyte hard drives are available to the consumer at a reasonable price which would store 4 million pictures. You'd have to take, or download, 274 pictures every day for 40 years to hit that number. In the physical world, we only have so much space and it doesn't really grow in the same sense. Think about your kitchen cabinets. You can only fit a certain number of cups and bowls inside your kitchen cabinets. Once you run out of space, you're out. You can maybe reorganize the cups and bowls in such a way that you might fit a few more, but the capacity is still the same. You can make the cups and bowls smaller but, at some point, they will no longer be useful for holding food and drink. You can make more cabinets but then you will need more physical space to do so.

In the digital world, we are somehow able to continuously fit more things into less physical space. There are now flash drives with the capacity of 1 Terabyte that are the size of a human thumb. More, we can actually copy the things that we have and place them in as many locations as we want. I can have two thousand pictures that I copy and put on a computer, a separate laptop, a hard drive, a USB drive, on Facebook, Google Drive, etc. If I lose my laptop, I do not lose my pictures. How is it, that the exact same picture can at the same time be in my laptop, a different computer, also across the country sitting on Facebook and Google's hard drives, and in my pocket? Why do we take for granted the fact that you can carry around a cell phone that is able to literally pull a movie out of thin air, enabling you to choose from a remote library of movies to watch anytime you want? How is it possible that thousands, or millions, of people can stream the exact same movie to various locations across the world at the exact same time? Is it so hard to believe that if there was a God, he could somehow make himself present in any and every place at the exact same time? Netflix is able to do it. I understand the historical issue with understanding God's relationship to space and time. The traits of God seemed so unfathomable. When people began to study space and time in a scientific way, they may have begun to feel as though the idea of God was so far out of the reach of material science that it could not even be sniffed, must less touched. I think of it like a man who is considering building a 500-foot tower. He can imagine the idea of the tower, but he gets overwhelmed with the thought of how he might even begin to build such a tower on his own. With no engineering experience, no industrial tools, he would think it so impossible that it's not even worth trying. However, with the right education, the right machinery and tools, and a team of competent people, he would have a different level of confidence.

I believe that the development of modern technology has helped us come to a better understanding of God. For a long time, it has been challenging our traditional notions of space and time. If you've ever played modern video games, you'll know what I'm talking about. The manipulation of space inside a video game is so fascinating that it keeps kids (and adults) entertained for hours on end. While playing a video game, the user is able to explore an entire world without physically leaving their living room. In some games, like Grand Theft Auto, the map of the video game corresponds pretty accurately to real locations like Los Angeles, California. A user can run through the city, or neighborhoods, or get in a car and drive on the highway to another part of town. What's fascinating to me is that this concept is no longer fascinating to most people. It's usually great fun for gamers, but there are many games that enable a user to explore a large map that gamers do not consider fun. We take for granted the fact that, for hours, we are sitting down, pushing buttons, and looking at a thin piece of glass that is covered by a plastic and metal frame. All the while, our minds are exploring vast cities and worlds. Some games, like Minecraft, allow you to build your own village or city. You can fly from one part of the world to another part and discover castles, islands, and animals. I've seen people build cruise ships, and roller coasters, and giant statues and I've walked around them like a tourist, looking up and down to see different parts of their creation. But did I really do any of those things? I mean, I was sitting down during all of that and my screen is only sixty inches in diagonal length. How could I explore a world, walk around, or build a castle that

I could walk through, with only sixty inches of space? Did I really look up when I was trying to view the top of some statue? My head didn't change position. We have merged our language which describes physical motion and spatial relations with how we describe our interaction with technology. We say that we have 500 Gigabytes of free "space" because that's how we see it. We can put stuff there like pictures, videos, documents, games, etc. So, it is space. But again, we are taking for granted that we are not physically putting a 5x7 picture into a space that can contain a 5x7 picture. I don't know anyone with a smartphone that's large enough to physically store a 5x7 picture, yet no one would question that their smartphone has the space store a photo of that size.

Many of us fail to recognize how much humanity's notion of space has changed over time because it's now so normal to us. It's normal experience to be handed an entire virtual world in the physical space of a compact disc (CD). It was not normal—and therefore probably would have been called magic— two thousand years ago. Those who are avid gamers might try to call me out on the fact that games don't fit on CDs anymore since we have Blu-Ray, and now 4K Blu-Ray, but I actually said the physical size of a CD which is the same, so, booyah. While the Theology of God's Omnipresence is not exactly the same, this example shows how some things are not bound by the traditional physical space restrictions we try to place on God. Many people see technology and believe there are no bounds but then try to place every kind of bound on the idea of God. It is a little Dr. Jekyll and Mr. Hyde, on one hand, to say, "I built a giant castle with a dungeon, fifty bedrooms, four dining areas, an indoor gym, and an Olympic size pool... in my 60 inch TV" and then, on the other hand, believe that the idea that God isn't bound by space and time is nonsense. Neither one makes sense when we measure by our own understanding of the limits of space and time. As we continue to innovate in technology, and push the boundaries of those limits, we expand our minds to stretch beyond those boundaries we once had. Imagine Bill and Ted took you into their time machine and you visited the smartest, and most advanced, people on the planet five hundred years ago. Imagine you told them that you could capture their image—like a painting, but one that is painted instantly—and then send their picture into space and then back down to a location across the ocean in under an hour. What would they say after they laughed at you? What if you didn't have a smartphone or a satellite but you could demonstrate a few things that would make your proposal sound more realistic. Let's say, you brought with you a walkie-talkie and you showed them that you could deliver your voice from one radio to another fifty feet away. Let's say that you also brought a disposable camera and you took a picture and developed the picture in a dark room. While you have not demonstrated that you can send a photo of them into space and across the ocean, you have demonstrated that they are severely lacking in their capacity to determine what is, or is not, possible. It is as ridiculous to laugh off the possibility that, one, God exists and, two, God is outside of space. We simply don't have the capacity to make that determination with any sort of authority.

That God can be outside time is also a thing that is not so hard to believe because of technology like cameras. Before computers ever came along, cameras were a technology that challenged our concept of space and time. The introduction of photography allowed humans to, in a way, go back to a snapshot in time and experience that moment in time in a real way. Video is an even greater testament to that idea. When we watch a video of ourselves, or a loved one, we are able to experience the sounds and sights of the moment. We can hear the words "I love you" from someone who passed away years ago. We can enjoy their laugh again; remember what it is like to see their smile and be in their company. Movies allow us to experience different time periods and witness the style, accents, and personality of people who lived and died before us. Part of the reason I think that we take these things for granted is because they evolved over generations and many of us have grown up with these technologies in existence since our childhood. In many ways, we have been lifting the restrictions of time and space that we believed we had on us for hundreds of years. Before people could take a snapshot on a camera, they could paint a person's image and capture something of that moment in time by painting their experience. While the camera has become more advanced and more accessible to the average person, it has also been around for generations. So, it may not seem as magical to us, but the more you ponder the fact that a camera can somehow replicate the exact shape of someone's face, the precise color of their eyes and the beauty of their smile, the more you become struck by the wonder of it all. How can a device, physically, recreate all of those shapes and colors and shadows in an instant without ever touching the object? All one has to do is click a button and an instant later they have a portion, or a replica, of time and space that they can put in their pocket and view any time. One could therefore go and "visit" some other time of history when they chose to do so. Eventually Thomas Edison had the bright idea (I know you see what I did there) that he could simultaneously snap many moments in time and then replay those moments. And thus, motion pictures were born. Those who take movies for granted may not realize that what is really happening is that you are being shown millions of pictures in succession at such a precise speed that it feels like you are actually experiencing that moment of time. In other words, if a camera is able to snap pictures fast enough, then it can track movements with great detail. Then, when the pictures are replayed—if they are replayed at the right speed—it would feel the same as we would expect we would experience it if we were watching the scene unfold live in front of us. Replay the pictures too slowly and suddenly you enter a time warp where people are in slow motion. Replay the pictures too quickly and people move as fast as squirrels. What's amazing is that we can actually intentionally manipulate these settings to, say, slow down the replaying of pictures in order to watch a scene unfold slowly in order to better understand it. This is very much a manipulation of space and time! We are taking real events that occurred in real time and saving them off for use somewhere else. We then are bending time to replay the event slower or faster or just from the beginning to the end again. All this happens in a different place than where the recorded event occurred. We can then clone that event so that people all around the world can view it at the same time from wherever they may be.

Side Note: I'm convinced that in the not so far future people will figure out how to reproduce not only the sights, and sounds, of a scene, but also the smells of the moment by capturing the data of the environments odors and reproducing them at will. Trust me. One day, Smell-O-Vision will be a reality.

Double Side Note: I just looked it up. Scientists have already created the sensors that can detect odor patterns. We're closer to Smell-O-Vision than you think.

It seems to me that if there were a God who created humans it wouldn't be hard to believe that he was not bound by space and time since humans—who are a part of creation—have the ability to nearly step outside of space and time. If a God created time and space, it is goofy to put him inside the restrictions of his creation when we constantly seek to push the boundaries ourselves. These examples don't prove that God exists, but they do help to provide evidence for what people of Theology have always said: God can place himself within space and time while not being bound by space and time. There is a difference in what we can do. It's analogous, so I recognize that we haven't proven anything. All I'm making the case for is the capacity for belief in things that are difficult concepts to comprehend. I think we know deep somewhere in our hearts that there is something, or someone, outside of space and time and therefore humanity finds ways to satisfy that curiosity. The advances in technology, and our manipulation of space and time, provide very good evidence for a belief in a supernatural God outside of space and time. No longer should it sound crazy to the modern person who enjoys watching someone get hit in the face with a soccer ball in super slow motion. You, my friend, have experienced a period of time warped in a way that no one five hundred years ago could have imagined. If you watched it more than once, then you have experienced that same moment of time, multiple times. By all traditional reason, your mind should be blown right now. You're lucky if you are still able to read this because your mind is still in one piece.

While those examples help counter the stumbling block of belief in God's nature, there are also positive examples in technology that provide evidence for God's existence. As we discussed before, the makeup of a computer is four basic components: input, processor, memory, and output. On a basic level, when you press a key on your keyboard (input), an electrical signal is sent along a wire to the brain of the machine. This brain is also known as the CPU or central processing unit. The CPU will first translate the signal and then do something with it. That's the processing part. Memory, or storage, is necessary for the computer to store data that it will need to access immediately while processing (known as Random Access Memory or RAM) or at a later point in time (in which case it is placed on the Hard Drive). RAM storage is volatile meaning that it is not intended for long-term storage. Once it loses power the data is gone. If you need to store something long term, the CPU would place it on the hard drive where it will persist even if it loses power. Lastly, the computer must be able to produce some sort of output. The CPU may pull memory off of the hard drive and translate it into code that can be transmitted to an external device like a monitor or speakers. If it sends the data through electrical signals encoded for the monitor to the speakers, the speakers will not know how to handle it. They will probably pulse terrible, thrashing, static. The output device must be able to handle the data sent to it and so it must be translated properly from its resting form by the CPU to enable that communication. The human body works very much the same. Again, on a basic level, you can see that the human body has all the same components. Our sensors provide us input through touch, smell, sight, and sound, and taste. Each sensor has peripherals that are connected to the body through, essentially, wires (i.e. nerves) that go to the brain. When you touch your keyboard on your computer, an electrical signal is sent across your nerves from your finger to your brain where your brain translates the signal to understand what kind of input you are taking in. The brain, like a computer, utilizes short term and long-term memory.

When I give a talk on this subject live, I usually ask people to humor me with a little exercise. If you close your eyes right now, can you try and remember what is around you? Can you remember the color of your curtains, or the design? Do you remember the color shirt your co-worker was wearing earlier today, or yesterday? Before you close your eyes, look at your hands, for as long as you want. Can you recreate all the lines in your hands when your eyes are closed? Obviously if you know what you are trying to do before you close your eyes you might be able to get closer, but you'll still be far away from perfect. Here's the crazy part. When your eyes are open you are constantly taking in input and processing data. Your mind is processing different colors, depth, length, the number of certain things like your fingers, movement, etc. All that information is stored immediately in your short-term memory. Like RAM, once you close your eyes your mind thinks it no longer needs this information and so it dumps it all. If you need to remember something long-term, then your brain puts that information into long-term memory. If a punch is coming right to your face, you are not going to be processing that information and throwing it into long-term memory. The movement of a fist coming toward your face is information needed immediately and your brain knows how to prioritize it. Your brain must then be able to translate that information into something that an external component of your body can understand. If your eyes see a fist coming straight for your nose, the brain can (and should) output signals to your legs to step back and to your arms and hands to rise and attempt to cover your face. Sound is another example. You might hear a song through the input of your ears, store that song in your long-term memory, and then days later your brain turns that song into output and you start singing. When we speak, we are resonating sound at a particular frequency and mimicking a common, or standard, set of sounds that everyone else subscribes to. Language is nothing more than a group of people agreeing to certain sounds being ascribed to certain concepts. Computer speakers can vibrate at the same frequency as a person's voice (or extremely close). Isn't that fascinating and at the same time sort of terrifying? I see this as great evidence of God's existence because the human body came along long before the computer. While I'm not equating the two, I am saying that the logic is very much the same.

Logic is logic is logic. No matter where it is expressed. Computers did not invent logic. They are

a manipulation of logic—a representation of logic –
in very creative form. The logic of computers has
always existed. Technically, smartphones could
have been produced when Jesus was walking the
earth two thousand years ago. If someone simply
had the knowledge to build a computer, they could
have. There was no material that was created or law
of physics that changed in order to enable
computers' existence. Here's what we can gather
from that fact: we know that computers are
intelligent devices and we know intelligent devices
must be designed and created. It's hard to imagine
taking someone seriously who walks upon a
computer and believes it to be a product of a perfect
accident of natural chemical reactions and
materials falling into the right place at precisely the
right time. It's easy to believe that computers were
designed by someone, or by a set of people, because
it takes intelligence to implement intelligence. It's
the same reason that you can tell sloppy, lazy, work
from careful and methodical works of passion. The
creation very much reveals the creator and when
you study some creation you, in turn, are learning
about the creator. This is the reason that I believe
people who really understand computers and
technology have a deeper understanding of God
than they realize. The logic implemented in
computers comes from the mind of God. God is
rational and logical. God created the material world
and therefore it is grounded in the elements that
are expressed through his movements.

Here's an example. If you found out that Michael Bay was going to be producing the next Rocky Balboa movie, you'd expect that Rocky would be running away from explosions left and right. I don't know Michael Bay personally, but I would bet he loves fireworks and explosions. Who he is comes out in his work. Michael Bay movies are recognizable for their large number of explosions that occur throughout the movie. Who God is comes out in his work. Thus, God is ordered and logical. Material creation was created in an orderly and logical manner, so it has the genetic makeup of logic. Unlike space and time, which are material, logic is immaterial truth. It doesn't have to "be" somewhere. It can exist in the mind because it's the conceptual structure of creation. This is why cameras can capture a moment of time, because they are taking in the logical aspects represented in the material: shapes, colors, depth, lighting, etc. They are taking in the "data" of life; the information that describes life. They then are "saving" the logical information, or data, in a pattern that can be later reinterpreted and produced as output. Logic can be then re-presented in many different expressions. For example, we can take that photo and resize it to be smaller or larger. Similar to an analogy. We take a truth and condense it by keeping all the important things and shedding the things that are not absolutely important. The more you shed the less the analogy resembles the thing it is trying to represent. In the same way, the smaller you make a photo, the more the important aspects of the information are lost, and the picture becomes distorted and pixelated. In a technical sense this would be called compression. If you ever have zipped up a file, you have performed this action. Technological devices are like devices that produce analogies of reality. Better devices will generally produce more accurate representations but that is essentially what they do for us. They re-present to us truths that we often want to dive deeper into and spend more time in. Much like a philosophy teacher who spends an hour lecture on what it means to exist, one might look at a snapshot of a photo for a long period of time trying to reflect on every part of that moment. It's not the moment itself that you are reflecting on. It's a re-presentation of the moment. You're reflecting on the logical concepts, or the data of the moment through a particular interpretation or expression. Think about it, what you are actually looking at is a series of colored ink splotches that are placed in such a way that the ink resembles (or re-assembles) the original moment. This is all made possible through converting (processing) real life experience (input) down into mathematical patterns, or equations, that then can sit somewhere (memory) in the form of ones and zeros and then be converted back up into an expression (output). All life follows this same basic logic! The idea is that all of real life is a mathematical language. So, when the Wachowski brothers had Neo seeing the Matrix world in streaming numbers—they were spot on! Well, sort of. The idea was right.

This language of logic is evident in that everything in the world can be represented through numbers using the computer system of input, processing, memory, and output. Even the human body is bound to this system. The question is, who created the system? Is it possible that an intelligent, logical, system could come to existence through illogical means (i.e. a total accident or chaos)? Other than the question of God's existence, where else do you see that sort of logic applied (that complete accidental chaos forms perfect logical systems)? There is much greater evidence to support the idea that there was a grand creator who designed the logical system that we humans have recognized and become participators in. I believe the language of logic, or math, is the language of God in communicating truth and order to the world. Interestingly, one day I discovered a great parallel between binary math, which is the basis for computer technology, and the Christian faith. We've already covered how binary is a system, or set, of only two numbers (1 and 0, in case you forgot) and those two numbers can be placed in patterns that represent everything on computers. A very, very, long string of ones and zeros is the resting form of your favorite movie or video game.

What's really fascinating and overlooked here is that there is really only one number in this set. The number one. Zero is not actually a number. It's a placeholder and the representation of the absence of a number. It's the absence of the number one. Zero is what you have when you have one of something, and then you take that thing away and now you have zero of that thing. So, zero is really describing what you don't have; something that's missing. In computer terms, the two numbers usually represent two states of current: true (1) or false (0). All things represented in computers can be reduced all the way down to a pattern of true or false states. If it seems like I'm repeating myself, I definitely am. You're not crazy. But I'm repeating myself because it's such an important and fascinating concept. Two simple states of representation: the existence of a thing or the non-existence of a thing, true or false, one or zero, are the building blocks of all computer technology. All things on computer devices have this at their core. Christian theology, similarly, finds its basis—its most reduced form—in two things: good and evil. And ancient Christian theology has always taught that evil is not its own thing. Evil, like zero, is described as the absence of good. It is not a thing we can point to, it's what we recognize when the good we expect is missing. All theology, at its core, is based on the foundation of good and evil, God or no God. Heaven is described as the place where God is. Hell is where God is not. All the complexities of theology must be built on top of this foundation just as all the complexities of computer science are built upon binary. I never set out to prove to you that God exists, just to show you that it's reasonable to believe he does. The evidence is strong and anyone who ridicules someone for belief in God, saying that it's not rooted in science or reason, is either being dishonest or has not come across the evidence. Again, this book does not set out to prove God's existence because, as I said before, proof seeks to remove choice. If I try to bury you with a mountain of irrefutable proofs, then I'm not respecting your ability to choose to believe. If I provide you reasonable evidence, the choice remains yours. At some point, like a detective, you must take all the evidence and make a decision on what it means. You may be wondering why God would leave it up to you. Why not just appear in front of you now? Then you'd know he exists, and you'd never have to worry about whether you made the right choice or the wrong choice. Great question. Let's keep digging.

FAITH AND REASON

I remember my first IT job out of college. I had just graduated with a bachelor's degree in Computer Science from the University of Texas at San Antonio. My grade point average was terrible because I often fell asleep in class and did not turn in assignments. The first two years of college, I worked almost full time and partied the rest of the time. I only passed many of my classes because I was either pretty familiar with the material—since I had been programming since 6th grade—or because I went to the library the day before the test and read all the chapters that the test covered. Usually this meant that I did not sleep at all that night. I would walk into the classroom, take the test, go back to my dorm and sleep for however many hours I could until I had to go to work. I was an idiot. I then met Dianne, whom I fell in love with and mid-way through my junior year, I realized I needed to change some things if I wanted to be marriage material. I started working hard to do really well in my classes, but I now had a new distraction that I poured a lot of time and effort into. And her name was Dianne. But that's not what this book is about. An extra semester on top of four years and a little grace sprinkled on top of me made me a college graduate. When I was interviewed for a job as a software developer, I had to make a case for why I would be a good hire despite my low GPA. In my interview, I basically laid out the same story of admitting my faults and not making any excuses for the bad decision to keep school at such a low priority. But I also talked about my conversion to wanting to become a better person, start a family, and be a great example to my future kids. I also mentioned that I had a desire to help other young people avoid my same mistakes and so I volunteered often to provide that message to young people. Obviously, I mentioned that my faith was a big part of my character. I still remember the perplexed look that my soon-to-be new boss gave me when he said, "That's weird, I didn't expect you to be really religious since you're a computer science grad". That was not the only time I heard something like that and, to be fair, I didn't know anyone in my computer science classes that was deeply religious. So, there's some truth there that people generally believe that computer science and faith do not jive. But why is that?

This has been an illogically brewing war for some time. I have seen so many movies where the protagonist comes to some stumbling block to belief in something. Rather than attempt to reason through the situation, they get some advice that they "just need to have faith". This is especially the case in romance movies. Some female is often like "Alex is the perfect guy. He's reliable. He's sweet, but he's safe. He doesn't take any chances. Mario is wild, and crazy, and fun, but he can be selfish and hates the idea of marriage". Then, some bad friend gives her bad advice like, "You just gotta have faith that whoever you choose, it'll work out". And then, stupidly, the first female is like "You're right. What would I do without you?" and then they hug it out.

This incorrect use of the word faith is why so many people believe it to be incompatible with reason. It's so often used as a plug for the bleeding hole of insanity. If you can't make sense of something, you gotta have faith. Didn't study for your test? Just have faith it's all gonna work out in the end. Don't have a good reason for why you believe in God? Well, good thing you have faith, so you don't need a reason.

The average person's knowledge of math, science, and history grows because of educational institutions that instill that knowledge in people starting at an early age. I think it used to be far more common that many young children were also brought up being taught about their faith. I mean, with a serious amount of time investment. My guess is that once educational institutions became public offerings on behalf of the government, religious instruction became less common as it can't be a part of required public school curriculum. Going back to the industrial revolution, when it became more commonplace to have both parents working full-time, daycares and public schools became a necessity. Students, spending most of their waking day with teachers at public schools, began to receive a great deal of influence from their teachers and less instruction from their parents. This is still the case today. I don't mean that parents back then, or today, ignored their duties to teach their kids lessons. Still, logically, there is far less time and energy in the evenings after a full day of work. It's also illogical to believe that someone who spends eight hours a day with children won't have significant influence on the way they see the world.

For clarification, I'm not saying that people today are not as smart - in regard to religion - as they were in the past. The average person now has access to the world's best resources to learn about their faith. What I am saying is that a larger number of children grow up without having significant and prolonged exposure to religion from the inside. Let me make an example of what I mean. We now understand far more, scientifically, about food and growing food than we ever have before. Anyone can do a quick google search to learn about which foods have the most vitamins, how to grow tomatoes, and various other things. However, the average person does not know how to grow food. They never have. Ask someone about growing food and they can probably give some basic information based on natural logic, but they will not have the hands-on experience that is required to truly know a thing. I know a lot of information about famous people and movie stars that I have never met. I don't really know them. No matter how many interviews, or movies, of them that I watch I still won't know them. I have to spend time with them. It can't be that I met them once or even that I casually see them at work or at church. To really, truly, know someone I have to spend significant time with them. This is like the difference between people who might have more memorized, and abstract, information about something, or someone, and those who might not be able to communicate their experience in abstract language but who truly understand the subject matter intimately. The way the latter person demonstrates their knowledge of the subject matter is not by speaking but, rather, by doing.

A farmer may not be able to explain why certain crops grow better next to other crops. He might not know the chemical reactions taking place, or anything about the evolution of the species, but he has been doing it for decades because he knows it works. Let's imagine that some scientists, fresh out of college, study the crops in a lab and conclude that the farmer is wrong. They then try an experiment on his land where they place both crops far apart and then next to each other. Miraculously, the crops that are next to each other grow faster and healthier. It seems the farmer was right. After further study, the scientists conclude that one set of crops did better because the crops next to it provided a natural element that fought off pests and infections. The acid in the soil caused a chemical reaction in one crop and the other crop benefited by being in close proximity. The farmer was right, but he had no idea why. This is a simple example and I recognize that the scientists could have successfully reproduced the results in their lab if they were using the correct soil. However, this kind of thing can happen to any scientist who has never had any experience growing crops and therefore does not know the importance of certain types of soil to certain types of crops. Even for those who do have some experience, or are expert scientists, they can still miss an important detail because it's not a detail that has been ingrained in their mind since childhood. In this sort of scenario, I'm not degrading science, logic, or reason. The farmer may have been right but without knowing why he was right he cannot reproduce his results elsewhere. That means that he might have just been lucky with the soil. Imagine that he tries to help his cousin grow the same crops on a different piece of land. The farmer may insist to his cousin that he just needs to put those crops together and it is guaranteed to work. However, his cousin will be unsuccessful due to the first farmers ignorance that the soil plays a big part in his own success.

Here's the takeaway: science on its own can be ignorant to great truth's because of lack of significant intimacy with the material. Experience on its own can be ignorant to great truths because of a lack of cognitive, or scientific, experimentation. Both illuminate each other and together they find those great truths. For science to do so it must come into the environment to explore and experiment. It cannot do so solely from the outside. Likewise, the person with hands-on experience must step out of the box of doing and into the scientific approach of thinking, measuring, experimenting, and theorizing. The Christian faith is not something that can be understood simply by reading books. It requires hands-on experience to understand. At the very least, it requires spending significant time with someone who has valuable experience.

I think many people have great confidence in science because they are in constant contact with people who teach it, or promote it, from an early age. The importance of learning about material and social sciences is ingrained in us in each time we are punished for not giving our full attention in the classroom and through each reward for "good grades" when we demonstrate we have retained the instruction. It's not too different from religious indoctrination. Kids do not really have a choice in that they must go to school, respect their teachers and learn. If they cannot demonstrate their ability to grasp and retain the material, they will not be eligible to have fun in school sponsored activities like sports and school clubs. Even outside of school, employers may not want to hire someone who cannot demonstrate they received good grades in school. Parents who do not take their kids to school, or school them appropriately, can get into serious legal trouble. These are all reasons that it's insanely uncommon to meet someone who never went to school or thinks elementary, middle school, or high school is completely useless. It's widely recognized that education in the sciences is high on the totem pole of important things in life. Don't get me wrong. I fully support this effort to better educate ourselves and our children. I'm just demonstrating how it's indoctrinated in all of us from an early age. Because there is a lack of similar indoctrination in, and emphasis on the importance of, Christian Theology

(I recognize there is some in church organizations but there is nowhere near the strict enforcement that we see in material sciences), there is in many minds a natural imbalance of weight in regard to faith and reason. As we come to be more educated, and understand the value of that education, it is tempting to devalue the aspect of faith. Faith is like the underrated basketball player that helps you win a championship. He's completely necessary but he may not have all the flash that fills the seats. Without him the team doesn't win a championship. With him, the team wins a championship, but he never gets the credit he deserves.

It's important that we make sure our terms are understood. Faith is not the opposite of reason. The best synonym for faith is trust. Trust is not the opposite of reason. Trust, or faith, is a decision that we make based on what we can reason. It's the action we take based on the information we have available to us. Faith, in the religious sense, takes you beyond reason. It takes you further than reason because faith is able to go where reason cannot. To use an earlier example, you may feel that you know a celebrity because you know a lot about them. You've watched their movies, seen their interviews, and read all of their twitter posts. By all accounts, you deem them to be an awesome person and you would love to befriend that celebrity. Imagine you meet that person, and they ask if you have $500 that they can borrow. He says he forgot his wallet at home and wants to buy this great jacket. He promises to pay you back the following week. What do you do? He promised to pay you back, so the right answer is to give him the money, of course. Except, not really. Don't be like the bad friend from our earlier example who tells you to not think things through and instead, "just have faith". Giving that person $500 because you saw a few of his interviews is illogical. This is an example of where faith can be present without reason. Faith without reasonable evidence can make people unreasonable. "Just have faith" is never a valid excuse for a person who is being unreasonable.

Let's change the scenario a little bit. Imagine that your best friend is asking you to borrow $500. You have been best friends for over ten years. Your friend has proven to you time and time again to be the most trustworthy person you know. At one point your friend even loaned you $200 for rent during your college years. Your friend asks for $500 to help him make ends meet for the month. He doesn't have a timeline of when he can pay you back, but he promises he will pay you back at some point. You decide not to loan him the money as you do not trust he will be able to repay you. You simply don't have any information showing that he will be able to make enough money to meet his obligations and pay you back. This is an example of reason without faith. Reason without faith can cause stunted relationships due to trust issues. In this scenario, reason alone cannot bring a person to give a friend $500. When considering only the information available, the numbers don't add up. It would require faith and trust in the credibility of that person that they will get to a better place where they will then be able to pay you back. Your faith in the integrity of the person can give you comfort that they will recognize the value of your act of sacrificing $500 and therefore will not take it for granted. This attitude doesn't come from having faith in something that you have no good reason for. You have over ten years of instances where this person has demonstrated integrity, trustworthiness, and commitment to your friendship. You might think, "I know this guy. I know he will pay me back". But in reality, you don't know your friend will pay you back. You can't know the future with any certainty. You choose to believe that he will because that is what all the evidence points to. You have faith in your friend because of the evidence available to you. For all you know, your friend could be a foreign spy and he's spent ten years gaining your trust for some devious plan. Asking for $500 could be a test of your level of trust. Faith is a decision to have trust in something that you can never know with complete certainty.

I like to think of the relationship between faith and reason like romance and logic. Relationships need to have both romance and logic to be successful. Romance without logic leads to the kind of relationships where people get married after only knowing each other for a week. Logic without romance leads to a dull relationship that lacks meaning and excitement. The two need each other and illuminate each other to bring out the best in each other. Very much like a marriage. A relationship filled with romance that is tempered by logic will allow the romance to continue to flourish and be sustained over time because it is built on sound principles. The logic in the relationship must also be injected with romance to ensure that the logic is aimed at greater meaning and reaching greater heights. You can't only crunch numbers to come to a decision on whether to loan your friend $500. Pure logic doesn't help you to understand the value of helping a friend in need. Pure logic won't push you to empathize with your friend and try to feel what they are going through. At the same time, pure romance can never look objectively at the information to decipher a really bad idea. They push and pull each other and create balance. They are like two legs that not only stabilize a person but also help them to walk forward.

Reason allows you to take in information and walk through the logical system of truths in an attempt to decide what to do or what to believe. Faith then jumps off that platform to allow you to reach even higher. Mathematics is an awesome example of this idea. When the US was in the race to send the first man into space, they used every kind of calculation that they knew of in order to anticipate what it would take for a successful launch. In addition, they had to try and account for every kind of known variable to have a plan of action for each kind of scenario. Still, they knew that there was a possibility that there were variables they didn't know about and situations that they may not have accounted for. At some point, reason could only take them so far. The act of sending a man into space required faith. Alan Shepard had to believe in the abilities of the scientists. He had to have confidence that they tried their best to make it a safe mission for him. He had to believe that they didn't make any major mistakes out of carelessness, exhaustion, or maliciousness. He had to have faith that traveling to space was possible and that he could handle the journey. Obviously, he heavily utilized his reasoning to come to his conclusion, but it was the act of faith that took the scientific theories and made them an inspirational story. Overcoming fear, falling in love, sacrificing for others, and becoming an inspirational leader are not a result of reason alone. Yet, so many people try to attribute them to reason only because of their great faith in science. Take a look at what is occurring in sports media. People are performing backflip analytics on players and teams in an attempt to anticipate who will win and to determine who the best players are. Old school players and analysts often look at these academics with disdain because they leave out other important player traits like grit, emotion, heart, and competitiveness.

There are some things that numbers cannot represent. While binary can absolutely represent every material reality, it cannot represent the immaterial realities like love, joy, anger, betrayal, sacrifice, or any kind of emotion. It can represent the logical elements that are a part of the expression of these realities. For example, a smile is a material representation of inner joy. A computer can represent the shapes and colors and shadows that makeup a picture of a person smiling. But those are just the outer elements of what we associate with the inner feeling that exists. We know these emotions exists and the only way we can demonstrate their existence is through the actions we associate with them. The problem is that the logical expressions that we see may not actually match the reality. Someone can be smiling even though they are actually very unhappy. The computer will then represent the incorrect reality as all it can observe and represent is the logical aspects of reality. In order to know how a person really feels, revelation is necessary. The person needs to reveal to you what they are feeling. Sometimes people may inadvertently reveal something when their emotion breaks out of their control and you can then reasonably deduce what they might be feeling. Still, they could be tricking us. There is only so much you can know about a person through reason alone. Revelation is an aspect of faith that you cannot verify using material scientific tools. I have been married for eight years and sometimes I feel like my wife is not exactly happy. I'll ask her if she's okay and she'll say she's fine. I'll squint my eyes and tilt my head as if I'm having trouble seeing the truth and ask her again. After eight years of marriage, I can tell you that there are times where I'm legitimately confused as to whether she's fine or not. Sometimes, it's obvious she's not fine. Other times, I have to give up asking her and then later she tells me the truth. This can be frustrating, but this tension is necessary for freedom to exist in a relationship.

If my wife was something that could be exactly measured using scientific tools, she would be something and not someone. If she was always scientifically predictable, there would be no mystery in the relationship. Further, there would be no need for trust. If my wife was only a matter of logic, like a computer, then she could be manipulated like a robot to my will. If I could know every single detail about her without her revealing anything to me, she wouldn't be able to hide anything and therefore she wouldn't be able to choose to be honest. Faith, like romance, allows you to go deeper into an intimate relationship with a person. It allows for mystery which provides the opportunity for trust. You don't have faith in things you know. You have faith in things that you can't know with 100% certainty. You cannot know a person in the way that you can know one plus one equals two. More, knowing that a person exists does not result in a deep and intimate relationship with them. Trust develops intimacy in a relationship and trust can only be affirmed when sound principles are tested and verified. Our reason provides us the ability to recognize the logical principles that make life good. For example, we believe people should not steal from each other. It's not good for an individual and it's destructive to a society. If we put our belongings in the custody of a friend, we are exercising trust. If they don't steal from us and they take care of our belongings, then our trust in them is affirmed. The principle we have reasoned would be good for us is affirmed as we experience the good of someone caring for our things. As we develop trust in a person, we then may begin to feel more comfortable with being vulnerable before them. We may decide to reveal things about ourselves that we don't tell anyone else. We might become more prone to ask this person for help. The more they protect our dignity the more trust we place in them. This is the definition of the act of religious faith: learning to exercise greater trust in God as we grow a deeper relationship with him.

Christianity and its theology are not simply an intellectual exercise. You cannot get to Heaven by passing an academic test. Complex theology, the bible, and writings of various scholars provide information about God to assist a person in actually coming to know God. But just like in a real relationship, listening to, and learning about, the other person can only take you so far. You must demonstrate your commitment in the relationship through acts of trust, or faith. And in order to do that, there must exist mystery in the relationship. Earlier I posed a question that, understandably, you might have wondered at some point. Why doesn't God just appear and let us know he exists? Well, God has appeared in multiple instances throughout history. Jesus walked the earth for 33 years. Of course, I recognize that no one has the ability to "verify" that claim in the way that they would want it verified. Still, people who ask this question haven't really thought through all the implications. In order to totally squash people's ability to doubt God's existence, he would have to appear to every individual. Because if he appeared to 50% of people on earth, the other 50% could still believe that those who claim to have seen God are lying or hallucinating. If he appeared once to everybody, then after some number of years, people would begin to doubt what they saw. A generation later, there would be people who never saw God appear and then they would doubt anyone ever saw God. If video was taken and became viral, the videos authenticity would be questioned. Even if it was believed, after time it would no longer retain its potency. This is not an untested theory. This sort of thing happens all the time. We see inspiring videos of people who do amazing things. We start to make resolutions to be better. Someone goes out of their way to do something truly amazing for us and we vow that we will always be loyal to them. We get into big trouble for doing something we knew to be wrong and promise ourselves to never do that thing again. No matter how inspired, joyful, angry, saddened, or desperate we may be in a moment, time usually changes how we remember that moment. Not that we might remember a sad moment instead as a happy one but that we no longer feel the emotions as strongly. Thus, it's easier to question whether our original feelings or observations were accurate. Think about one of your middle school, or high school, breakups. How heartbreaking was it? Did you think you would never be happy again?

I remember when I got the call offering me my first job as a computer programmer. I jumped up and down and started punching the air out of excitement. If I were to get a call today offering me the same job, I would turn it down with no hesitation. My salary at the time was less than half of what it is now. It's not that my feelings at the time weren't valid. They totally were but I can't resurrect the feelings I had that day. I have to give great effort to remember what that moment meant and to continue to be grateful for the people who helped me get that job. I have to make a conscious effort to recognize that my current career received great contributions from my time in that job. So, that job offer still should have great significance to me. If my old boss called me every day to remind me, I wouldn't have to remember, and I would feel like I was being manipulated. He would be correct in what he said but he would be wrong in his method to try and help me stay grateful to him for the opportunity he gave me. There's great value in the process of having to go through effort to be grateful, to remember the significance of moments, and to remember how people have helped you in the past. Here's what I'm getting at: in order for God to disarm the potential for disbelief, or doubt, he would have to persistently show himself to each person each time that they even began to doubt. He would have to persistently perform miracles that proved us wrong and made us look silly for doubting. We wouldn't be free people if God conducted himself like that. Imagine the person who is sitting at his computer late at night and is tempted to look at pornography. If Jesus popped up behind him and was like "Hey, just dropping by to say hi and make sure you remember that I exist". How would that change that person's decision to view pornography? What if you were the one on the computer and Jesus was in the room and was just like "Hey, you don't mind if I just sit here and read a book, do you? Don't mind me. Just keep doing what you're doing"? How would that change your behavior on Facebook and Twitter? I'd probably be like "No, that's fine Jesus, I was just about to read my bible anyway" even though I really wanted to watch some Netflix. The point is that God respects our freedom and he wants us to be able to make the choice to trust him. In order to be able to make the choice to trust him, the choice to not trust him must be available to us.

Remember our discussion on binary math, two things must exist for choice to be possible. This is why, in the book of Genesis, God created the tree of life and then told Adam and Eve not to eat it. He then "left them alone" so they had the opportunity to make the choice on their own. It wasn't a set up. It was an opportunity. The book of Genesis did not say that God would never give them the fruit of the tree of knowledge. The serpent tricked Adam and Eve by telling them that God did not want them to be like him (like God). We now know that God does desire to give us knowledge and everlasting life. The tree was meant to provide them the opportunity to exercise their freedom to choose to allow God to be the giver of that good fruit. Again, this was an exercise of freedom to teach them that trust is what builds the bridge to the desires of their heart. As a leader, I can tell you that this is brilliant strategy. God is absolutely logical. The only way to provide people the opportunity to succeed is to provide them the opportunity to fail. Micromanagers are frustrating because they are micromanaging in order to try and prevent any potential failure. They can't give you a task and give you the room to do something great because they have such an unhealthy fear that you might fail. Their fear of failure causes them to essentially bulldoze over others to nearly end up doing all the work themselves. This is not God. He is not a micromanager and we don't want a micromanager but it's often what we ask for when we wonder why God doesn't just appear or do more to stop evil in the world. God desires us to make a free choice to love him and he is so adamant about respecting our freedom that he allows us to choose evil. That's not to say he doesn't intervene and make himself known. The bible is full of miracles and there are many more reported throughout modern history. There are Catholic saints who died and whose bodies have remained incorruptible, meaning some or all of their bodies didn't decay. Many Christians, like myself, will swear to God's providence being at the wheel during many parts of our lives. While I believe I have found overwhelming logical, reasonable, evidence for God and the Christian faith, still, there remains great mystery. There are many things I don't understand and many things that can't yet be known in the way that we would want to know them. One might see that as frustrating, but I see that as exciting. That it will take the rest of my life to study, pray, discern, and experiment in my faith is not a deterrent to me. It's exhilarating! Life would be boring without mystery and challenges. If I had nothing to work for, no prize would feel worthwhile. So often, we ask for the easy way out to get rid of our struggles and we forget how good it feels to overcome those obstacles. We forget that the process of struggling through understanding something is what makes us experts. Michael Jordan once said, "I have missed over 9000 shots in my career. I've lost almost 300 games. Twenty-six times I've been trusted to take the game winning shot and missed. I've failed over, and over, and over again in my life. And that is why I succeed". Having faith that each failure can be a learning experience that gets one closer to his goal is what makes a champion. Thomas Edison failed to create a sustainable light bulb a thousand times before he got it right. He replied to a question about his failures with "I didn't fail 1,000 times. The light bulb was an invention with 1,000 steps". Hacking, forensics, programming and any other area of computer science requires the same sort of dedication to craft. It requires looking places no one else looks and trying things that no one else has tried. The reason hacking and cyber security is so fun is because it's difficult. There's no joy in the easy answers. There's no creativity and innovation without mystery and potential.

Part of the allure of computers and technology is the mystery of what is possible. As a programmer, you have the ability to explore potential by interacting with the code. The world is full of mystery and potential because God is a programmer and nature is his code. Nature contains the logic of the mind of God because it comes from God. Science helps us to gain a better understanding of the code, or the laws of nature, so that we can interact with it and even become creative ourselves. We covered this concept when talking about math so some of this will sound very similar but with a new flavor. Nature's laws are very much like the code of a computer program. They are a part of the design of creation and we interact with them as though they were computer code. Nature's laws aren't just firing off all the time and producing the same results. At times they work like methods that need to be invoked. Sometimes they require input, or variables, and they then produce results or output, based on the input. Each person will experience the laws differently depending on the type of invocation and the variables provided. For example, we are all experiencing gravity at all times. However, someone who is skydiving is experiencing gravity in a very different way than someone who is sitting in a chair. If we were to look at nature as computer code than we can predict what might happen when we introduce certain variables to certain methods. For example, what do you think would happen if someone were to jump off a fifty-foot cliff? Do you think they would fall down at a very high speed and probably die when

they hit the ground? If that's what you believe –
why are you so sure that will happen? We assume
and expect that the laws of nature will be consistent
and always act according to how they act right now.
We believe that because we have spent a great deal
of time studying how the "code" of nature works
and we trust that the code is reliable. We trust that
the code will not suddenly change on us. Our
confidence in material science comes from its
predictability. It allows us to maintain a feeling of
control over our lives. This is why, some people will
refuse to believe in anything unless it is "proven" by
science. But how confident can you be in your
ability to reason unless you first have faith that your
method of reasoning is correct? Human reason is
flawed and must be formed. People don't just
"know" the truth. Each morning, when I leave for
work, I use the key to turn the engine over and I
start driving. I assume that the vehicle is going to
work according to its design and the laws/code of
nature. I never wonder if the car is going to just
start accelerating out of control or flying up into the
sky. I don't know that those things couldn't ever
happen, but I have good reason to have faith that
they won't. I don't need to pop the hood of the car
and personally verify that every engine and
electrical component is working properly before I
drive to work each day. I'm not an expert in
mechanics or combustion so I have to make an act
of faith each time I drive because I don't know how
the car engine is going to behave. There are some
things I can verify, within reason. I can check the
fuel gauge, I can check the engine temperature, and

I can check the oil level. Still, I don't go manually check the oil every day. No one does. No one lives without some level of faith.

There is no amount of logical data that any person can obtain to allow them to conduct their lives without acts of faith. We put faith in things and we place faith in people. We recognize that this is also part of the design of life. Faith gives meaning to life just as interpreters give meaning to computer code. Ones and zeros mean nothing if they are not interpreted. See, God is not just a programmer. He is also a lover. He is romantic and that aspect of who he is comes out even in the realm of technology. Even computers are bound to a marriage-like relationship with each other. The internet only works if two computers enter into a common union built on trust. One computer takes some information and translates it into ones and zeros. That computer initiates the sending of that information over electrical pulses to the other. The information provides the potential for something, like a movie, or a game, or website. Another computer must be open to receive the information. When it receives the data, it then must use its own faculties to interpret the data. Two computers with different parts, and/or different operating systems, from different manufacturers may interpret things differently. That interpretation can be barely different or so dramatically different that the data is unreadable or unrecognizable. This is why some televisions, Rokus, and Fire Sticks can translate 4k content and some cannot. They must speak a common language in order to understand each other. They must be able to give meaning to the binary signals that they receive or all the data in the world is useless. The logic of binary code in movies is absolutely necessary. Remove it and there is no movie. However, without interpretation the logic has no meaning, no value, and no excitement. We don't watch movies in binary even though everything in the movie—the sights, the sounds, the emotions we might receive—is contained in the binary code. No matter what we might proclaim, we don't simply want the logic. We want the romance. We want meaning. We want the aspect of truth found by experimenting and exercising trust in the mystery that we cannot know. Much like when the US sent a man into space for the first time. We want to confirm the potential we hope for to find out if there is something more out there. The good news is that there is not only something more out there. There is some*one* more out there.

[CTRL+F] SEARCHING FOR MEANING

The search for meaning in our society is evident in the way that we often misuse technology. Waking up and checking our phones first thing in the morning; the excitement we get from a text message notification and the quickness with which we reach to read it. Constantly posting photos of ourselves on Facebook, and Instagram, sending shocking pictures through Snapchat; detailing our lives on Twitter and obsessively following the lives of others are common. I probably don't have to provide any more examples to communicate to you all the ways that people become obsessive with their use of technology. For many of us, it rules our lives. It's not all bad, of course, I love being able to stay in constant contact with family members who live a thousand miles away from me. Still, even I struggle with allowing technology to be constantly at the forefront of my mind. Why is that the case? Why is technology so easily addicting? Why do people feel a desire to live their lives online? Although it can be argued that technology can make matters worse, technology doesn't create human problems. What technology does do, is magnify human problems that already exist.

In fact, we should be grateful that technology is helping us to better understand the human condition. Selfishness, arrogance, narcissism, gossip, and nosiness existed before technology came along. Again, I'll grant that technology can often be a vehicle and an enabler for people to further develop bad habits. However, it more so amplifies the traits that we have. Selfish people act selfishly online and we just notice it more because their actions are public, can be viewed repeatedly, and are subject to commentary. Humble people tend to act humbly on social media for the most part. Of course, there are a great deal of people who hide behind fake names, or handles, and act differently than they do normally. Again, this is an example of technology actually helping to reveal people's true self. Who you are when you think no one is looking is who you really are. Social media, despite its many downfalls, has reminded us of the great truth that material science, and reason, is not the end game. People have an intense desire for love and attention. No matter who they are. We've seen hardcore rappers, million-dollar athletes, and the President of the United States turn to social media for approval, to complain, and to connect with others on a deeper level. I'm not a people pleaser. I actually prefer my space and I'm rarely interacting on social media. Still, I desire the love and affection of my family and fellowship of my friends. Addictions are usually the result of desperate searching. When we are hungry, we want to eat. When we go a long time without eating, we become starved. Desperate hunger with no food nearby will cause people to search in unusual places. I'll recycle an analogy from Christopher West's Theology of the Body presentations: if you are starving and have nothing to eat you might consider eating out of the dumpster. The result is that while you might satisfy the immediate hunger, you will inevitably end up sick. Our intense hunger for love and affection, for human interaction and communion, draws us to the internet where the world never sleeps. Here we find the opportunity to connect with others, to make friends, and to share ourselves. We desire to see and be seen. We don't just want to know others, we want them to know us. So, we follow their thoughts on Twitter, cycle through their pictures on Instagram and engage in conversation on Facebook. But that's not enough. We want them to follow us. We want them to be interested in our thoughts. We want them to look at our pictures and admire our image. We want to know that people are interested in us. One of the most fascinating things about social media is the trend of people making themselves vulnerable before the world and opening themselves to the potential that people might harm them.

Because we often settle on social media platforms for fake love, fake affection and disingenuous interaction, we receive only temporary satisfaction. The void remains, and, in fact, it grows. Much like an addiction to alcohol, drinking temporarily numbs the pain of heartache and quiets the noise but it doesn't last. Worse, one becomes immune to the alcohol after some time. So, in order to numb the pain again, more amounts of alcohol are needed. Alcohol becomes the savior from the pain and the attachment to alcohol only grows. Social media and technology can often numb the hunger we have for attention and provide us an outlet for the urge we have to share ourselves. Let me be clear, this is not inherently a bad thing. Alcohol is not inherently bad. I love the taste of beer and a good Jack and Coke often hits the spot. Red wine is amazing. Social media providing a platform for us to connect with others and share our inner thoughts is awesome. I'm a big proponent of technology and social media. The danger comes when those things become the only way that we can connect with others or the only way we can communicate our thoughts and feelings. The danger comes when we lay in bed next to our spouse and, instead of talking about our day, we are browsing Facebook to learn more about other peoples' day. Just like with alcohol, drinking alcohol can be a good and pleasurable thing. The danger comes when you can't find good and pleasure in anything without the use of alcohol. Imagine you were in the presence of an alcoholic and you asked about his father. The man becomes angry and reaches for a bottle of scotch and begins to pour, and drink, one cup after another as he tried to convince you that he couldn't care less about his father. The fact that the topic drove him to begin burying himself in the bosom of his alcohol would provide insight into the fact that he has a profound wound in his relationship with his father. Where people emphasize their addictions is where you can normally find their deepest wounds and hungers. The person who is constantly posting photos of themselves on Instagram or Facebook with arrogant captions is likely someone who has a severe struggle with self-confidence. The one who is constantly looking for compliments from strangers is probably the one who lacks compliments from the ones they desire them from the most. For example, their parents or spouse. The one who posts revealing photos of their body is most likely a person who desperately wants someone to desire them with an eternal passion that wouldn't end when their body changed.

At the root of it all, our addictions are grasps at love. They are grasps at knowledge, power, justice, faithfulness, and communion. They are grasps at something outside ourselves, something greater than ourselves. They are the result of us reaching out to grab something that we know deep down is promised to us. Greatness. Not just the Michael Jordan, or Michael Jackson, kind of greatness where you are widely recognized as the greatest in your profession or trade. I'm talking about great meaning. I'm talking about being an important part of a great story where our lives have meaning because they are a part of a bigger storyline that has great meaning. We are obsessed with stories which is why movies and video games are so popular. It's why musicians usually put some kind of storyline in their lyrics. A great movie cannot just contain dazzling special effects. It cannot just have a big-name actor or a big budget. Movies based on DC comics have proven this point. There must be a powerful story for the movie to have great success. Stories speak to us. They inspire us, they teach us, and they remind us of great truths. We need a main character with whom we can relate. We need to understand them, know what motivates them, know what angers them, and be aware of their strengths and weaknesses. We need to see adversity. They need to struggle. They need to be the most powerful character but also at some point in the movie be so vulnerable that we fear they may lose big. Then they need to somehow dig deep, push beyond their limits and become victorious. This foundational storyline is used in war movies, romance movies, comedy movies, superhero films, chick flicks, anime, and sports movies. Some of our favorite movies are the ones where we can imagine placing ourselves as the main character, either because we want to be like the main character or because we already are.

These powerful stories are beloved because they remind us of a calling that we all have to greatness. We are the main character of our own story and like the main character in a movie, we are searching for all the elements of the story. Like the character at the beginning, we may be unsure of who we are, or of our place in the world, so we search. And we search, and search. At times we stumble and fall. We have failed experiments and eureka moments. Some people figure it out and find success and happiness while others spend their whole lives stuck in a loop of sad endings. I liken it to people searching through e-books for the answers to test questions. Without understanding the question, one might find themselves using Ctrl+F to aimlessly search through the vast text for clues to the answer. If you're lucky and you have the right keyword you may stumble upon the answer. If the keyword you are searching for isn't narrow enough, you're likely going to end up finding random results that may or may not lead you anywhere closer to the answer. Often people have a habit of trying to Ctrl+F through material and just submit answers blindly in the hopes that they have accidentally stumbled upon the answer. The reality is that you need to understand the material to find the answers. You have to actually read through the books, listen to the class lectures, and spend time thinking about, and contemplating, the material. Then when asked a question, you can use Ctrl+F to search for keywords that will lead you to the right book and the right chapter. There are so many competing solutions that claim to offer the path to the meaning of life and happiness that it can become discouraging. If you decide to believe in God and the idea of life after death, which religion offers the true version of that God?

As described in an earlier chapter, the world is made up of signs and symbols. Sign and symbols, by definition, point to something beyond themselves. They image, or represent, something greater in a way that can be more easily digested. Language is a form of symbolism for communicating our inner thoughts. We say, or write, what is in our minds and hearts. The act of producing communications according to a widely agreed upon system of sounds (spoken language) or ink splotches (written language) is an expression of symbolism. Computers are built on the use of signs and symbols. To be a great programmer, or systems level hacker, you have to be intimately familiar with the use of pointers. An operating system, like Microsoft Windows, or Linux, uses addresses and pointers to manage memory. Anytime you input information into your computer, or try to open a file, the computer follows a series of pointers to locations in memory to find data and instructions. Like mathematical equations, pointers become placeholders for complex information.

Let's perform a simple example, take the equation: 1+1=2. We can make x a pointer that points to the number 1 (x=1). Therefore, the equation 1+x=2 is the same as 1+1=2 because it is leveraging the pointer x which points to the number 1. In the variable x is the capacity to "hold" something else that can be more complex than x. In this example, 1 is only a digit and it takes up the same amount of space as x so it's sort of silly to make x a pointer to 1. How about if we took the equation, 2+x=42 and x was a pointer to the equation 4y. If x=4y, then 2+(4y)=42. That would mean y is a pointer to 10 and the equation would be 2+(4*10)=42.

$$2+X = 42$$

$$4Y$$

$$10$$

$$4\,(10) = 40$$

This might seem silly but pointers are necessary because there may be multiple solutions to this equation. It may be that x points to 5y. Then 2+(5y)=42 would mean y points to 8. If x pointed to 22+y, then 2+(22+y)=42 would mean y pointed to 18. The only way to know what something is pointing to is to follow the trail. The pointer x might lead you to an equation that involves y. That might then lead you to where y contains another equation that contains a z pointer.

$$2+X = 42$$

$$(22+Y) \longleftarrow Y \dashrightarrow 3\,(Z) \qquad 18$$

$$Z \dashrightarrow 6$$

Computer hard drives are broken into sectors and clusters which are essentially sections of the hard drive that are associated with addresses. These addresses are important because it allows the computer to know "where" to go to retrieve data, or to place new data. Pointers are what a computer uses to know what current instruction it is executing, and which one is next. Computers cannot communicate, or function, without using pointers, signs, and symbols and neither can we. Therefore, the whole world's communication is in signs and symbols. But what are the signs and symbols and symbols ultimately pointing to? Walk with me for a second through this logic. Let's assume there is a God and that God is logical and ordered. If that God created the world, then we should be able to learn about him through studying his creation. Let's say that he wanted to communicate to us. If we are material beings, how would he communicate to us? We established earlier that he uses parables, or analogies. Essentially, he uses signs and symbols. If he wanted to communicate to us that beauty is good, he would make the natural function of beauty something that returns pleasure. Think about it. We usually perform the reverse communication. If you were to meet an alien and they asked you, "what is beauty?", you would point them to the sunset, or mountains. You might point them to a beautiful woman, or a famous painting. This is actually the reverse of how God communicates to us. If someone asked us, "what is beauty?", we would be pointing them to the signs. Really, the signs (woman, sunset, mountains, painting, etc.) are pointers to the concept of beauty. They contain within them the expression, that points to the ultimate source of beauty itself. Conclusion: if the logic of the world is the communication of God to us through signs and symbols, then we should be able to trace the path of that communication back from the destination to its source. In other words, if God was 42, but all we had were the symbols of 2+x. We should be able to follow the pointer from x to 4y, and then follow y to 10, and we'd end at 42. Turns out that "deep thought" — the computer from *Hitchhiker's Guide to the Galaxy* — was right. The answer to life is 42.

It would certainly make sense, as we saw in an earlier chapter, of why Jesus was constantly speaking in parables as he used the signs and symbols of the world to point greater truths. Remember that logic, is logic, is logic. Truth is transcendent and though there are many expressions of meaning in our lives, they all are bound to point to the same transcendental truths. Sacrifice is at the heart of the Christian faith, but sacrifice isn't just a truth in the Christian faith. Sacrifice is good in business, sports, relationships, and even the game of chess! Conversely, selfishness is bad for business, bad in sports, and devastating to relationships. Teamwork and community are vital to a business organization, to hacker groups, to charitable organizations, to sports teams, and dance teams as well as the Church. Faithfulness and loyalty, similarly, are staples to all of the above-mentioned areas of life. It makes sense, then, that we search in these areas of life for the meaning that we so desperately crave. People pour their entire lives into their jobs, sports, their relationships, and communities because they recognize that they are symbolic expressions of the greater truths that we seek. They are areas where we can live out the calling we have to greatness. It's possible to recognize truth without fully recognizing why we are attracted to it. If we recognize the signs and symbols in the world as signs and symbols, we can look to the direction in which they point. Looking out at the ocean, or up at the sky, can point us to contemplate the vastness of the world. It can remind us that there is a much greater force that is beyond our control. While we can interact with the sky and the ocean, we never have full control over them. I mentioned earlier that I believe that music points us to the desire we have for faithfulness and consistency. The rocking of a baby in a consistent pattern or the repetition of a song provides the confirmation to our expectations which provides comfort. That the base two numeral system is the lowest possible reduction of numerical systems is a symbol of the great truth that choice and free will are at the heart of truth and meaning. The fact that technology seeks to push past the limits of space and time reveal that we know there must be something or someone outside of space and time.

The explosion of social media and the internet show us that it is a universal human need to be in community with others, to be loved by others, to share ourselves—our thoughts and our emotions, our lives—with others. This is also true in the open source communities and hacker communities. No one likes to build and enjoy accomplishments alone. That we love to watch, and tell, great stories through books and movies and video games reveals that we recognize the call to a great story. Still, all of these things are like pieces. The greater question everyone must answer for themselves, is, do they belong to a puzzle or are they just random pieces with no greater structure? If they belong to a greater structure, a puzzle with a particular design where all these pieces fit together, then I think it's more than reasonable to believe that there is a designer. More, it would mean that this designer is logical, strategic, and structured and he designed these pieces to fit together to form something greater than the pieces can be individually. If this is the case, then our search for meaning in all of the puzzle pieces, the signs and symbols, may really be a search for the one who designed the pieces. Ready Player One is a recent movie by Steven Spielberg where people spend most of their time in a vast virtual reality (VR) environment. It's similar to the environment of the internet but everyone is sort of interacting "inside" the internet through the use of VR goggles. The storyline of the movie involves the creator of the virtual world leaving behind three Easter eggs, or keys, inside the virtual world before his death. Whoever finds all three Easter eggs would obtain the rights to the virtual world and so everyone is trying desperately to be the first to find the three keys. In order to find the keys, players must study the life of the creator to figure out where he would have hidden them and how they could obtain them. The main character is able to find the first key as he is heavily educated in the life of the creator. He has seen every video of the creator's life, much like many others inside the game. He has a vast amount of informational knowledge of the creator. Studying the life of the creator helps him to figure out where he would have left the Easter eggs, but the journey to find the Easter eggs is what provides him an intimate understanding of the creator. Each Easter egg that he finds illuminates the information he had about the creator as he is able to jump into the creator's environment and interact with his creation. He goes back and forth between the virtual world and the creator's archives because each one illuminates the other and make his journey more fruitful. He eventually uncovers the hidden keys which grants him a personal meeting with the creator's avatar inside the game. Interestingly, it's revealed to the main character, as he comes to understand the creator in a personal way, that all the signs and symbols inside the virtual world, and the journey to uncover the great truths (or the hidden keys) were intended to communicate a profound message. By the end of the journey, the main character recognized that the creator was trying to communicate to him that there was more to life than the game. Each hidden key moved along a path that pointed to the creator's desire for human interaction and marital love. The creator could have simply come out and made a statement as such, but he was a brilliant creator and he wanted people to learn that lesson through experience. People had been playing the game for years trying to find the Easter eggs and some had invested their entire lives in dedication to that mission. By the end of the game, everyone watched the main character with bated breath because as much as some just wanted to be in control of the game, the real fans wanted to know the truth. They were desperate for the revelation of the truth. They wanted to understand the creator through his revelation. Those people cheered when the main character uncovered the last key and met the creator.

The search to uncover the hidden truths in our world through all the signs and symbols of creation, similarly, is really a search to uncover the creator behind the creation. The desires that we all have that cause us to search so intensely for meaning—if we search diligently—will lead us to the truth that God shares the same desires for love and communion. His creation reveals that he has created a journey that will lead us to understanding him, but even more so it should lead us to communion with him. Like in Ready Player One, the Bible details the story of creation, the story of God's people, and the story of Jesus's life on earth. To uncover the hidden truths, we must be willing to study the life of God but also recognize that the study alone will not provide the sufficient insight. We must also pursue the pearl of great price through our actions, we must seek truth in the archives and in the world. We must learn the lessons that God set up for us in the framework of the physical world. In a sense, we must play the game to enter into a deeper understanding of, and communion with, the creator. The answers are not out of our reach. Each one of us has a unique storyline that is set up to teach us the profound lessons of life that lead to everlasting life.

The desires that you have for love, for human interaction, for joy, for touch, for vulnerability, to share your self, to see and be seen, to know and be known, are good. The desires you have for technology, for knowledge, to create and to program are awesome. They should be affirmed! They are signs that point you to the meaning of life and they can lead you to receiving eternal life. The Catholic Church teaches that in Heaven, we will become *like* God. We will live forever, and we will enjoy the fruits of divine life. To be clear, it doesn't teach that we will be God, or that we will be gods with our own little worlds of people who worship us. The Church teaches that we will be a part of the life of God and thus we will enjoy divine qualities such as everlasting life and joy. You see, the desire that Adam and Eve had for the fruit of the tree of knowledge was good. They had desire for good things and God absolutely wanted to give them the desires of their heart. The problem was not that they wanted to eat the fruit but how the obtained it. God provided them freedom by giving them choice. They made the choice to *take* the fruit instead of choosing to allow God to be the giver of the gift. They failed to trust that God would fill their hearts and fulfill their desire for truth. God is most definitely a programmer and it is evident in the logic of his creation, but he is also a lover. He wants us to have freedom of choice because he wants a relationship with us. In any relationship, love must be given. It can never be taken.

I fully recognize that, to someone who does not share the ideals of the Christian faith or did not grow up in a Christian home, this may sound ridiculous or just weird. From the outside it may seem odd but from the inside it makes perfect sense. I am a father. I love my children. To an outsider, it may seem weird that I can't stop kissing them. It may seem ridiculous that I have four children running around yelling, at times, or that my two-year old's smelly feet can sometimes smell good to me. Inside the family, the sounds, smells, and laughter create a bond that outsiders would have trouble understanding because they are unaware of all the memories, struggles, hopes, disappointments and trials that we have been through together. Logic appeals to us because it is a language that makes sense of the complex concepts of physical reality. We recognize that language and we desire to understand it. That's why music can convey emotions to us. We know a happy, upbeat, song when we hear it even if it has no lyrics. Similarly, we know a sad, or scary, tune even if it lacks scary words. How weird would it be if a serious movie played action music in the background of a sad scene? It would convey the wrong emotion to the audience and ruin the scene.

Music is highly logical, which is why notes played on instruments or through speakers can produce beautiful mathematical patterns. This phenomenon is called Cymatics and is demonstrated through the use of sand, or dust, on a surface that is vibrating from sound. The sound vibrates the surface and the sand moves into place to create amazing geometric shapes and patterns. Even music expresses the language of logic. Our desire to express and communicate in this language comes out in our desire to engineer. That's why it's common for us to build contraptions, break things, and take them apart to see what's on the inside. We want to manipulate, and interact with, the world around us. We know, somehow, that there is a greater truth in this language of logic and if we could just understand it, if we could learn how to speak it, we could become more than bystanders. We can become imitators, creators, initiators, and innovators. We can become like God our father who is the creator, the initiator, and the innovator. We have always been told that God is love and we strive to be like him in that respect. However, we also have a deep, built-in, desire to be like God in an intelligent, creative, and participatory way. We want to be involved and hands-on and come to know God through imitation of God. I had great admiration for my dad growing up and I wanted to imitate many of his awesome qualities. I often see my children imitating me and looking in awe at the things I can do (which are usually hilariously simple things like typing on the computer, shooting a basketball, or beating a level in a video game).

Computers and technology have allowed us to be like God, to create our own worlds, to step outside of time and space, to build intelligent devices that others can interact with to obtain joy and become educated. Technology is helping us to make great strides in understanding an aspect of God that has always attracted us. The question is, do you really want to know the truth? Are you willing to go to the lengths that it takes to dive deep into the ocean of uncertainty and exercise trust in God? I can tell you from experience, I have been on that journey for well over a decade. When I first began to search for greater meaning, when I began to dive deep into the Catholic faith to satisfy my thirst for truth, it was scary. I was overwhelmed by the depths of the faith. I was terrified, at times, at the cost I would endure. What would people think of me? What would I have to give up, or take on? These were things that I had to painfully consider once I began to exercise trust, or belief, that this faith might be true. This is what romance requires. This is the beauty of faith. You never know what will happen until you take the chance, jump out into the deep, and place your investment in a well calculated risk. It may cost you upfront dearly, but I promise you, the reward will fill every appetite and longing you have for meaning.

HELLO, WORLD!

Technology is not going away, nor should it. I want to start this chapter by making an appeal to the religious people who picked up this book to figure out how to connect with people who love technology. It's up to us who are involved inside the circle of faith to be willing to go into depths of technology and recognize those who are thirsting for truth. Those who believe that technology is evil, or that it is a tool of the devil, have been swindled. They are like the puritans who believed that sex was more dangerous than any good it could provide. They believed that a woman who showed her ankle was scandalizing men because she was being too provocative. They believed that the answer was to demonize sex and the hunger that people felt. They shamed anyone who demonstrated their thirst for sexual love.

Hugh Hefner once said in an interview that his puritanical childhood led to him create Playboy[12]. His antidote was to break the shackles of sexual repression and then promote the opposite extreme: indulgence in sexual desire. I understand the fear of the possibilities that technology enables. I wonder if, at some point, people will be able to hack the brain by creating artificial sensory input and tricking the brain into seeing things that are not there or hearing things that didn't come from sound waves. I know right now drugs sort of hack the body and can cause hallucinations, but I mean hacking in that one day we might be able to provide specific data that manipulates the brain into believing it sees something that we created or bypassing the eardrums to put a song into someone's head. Scarier, what if someone figures out how to place a specific thought in someone's mind using binary signals and some technological transmission medium not yet invented? I was told recently that Dr. Gregory Carpenter successfully demonstrated the use of nanorobotics to control human muscle movements at INFOWARCON. In these kind of applications, and their worst use cases, is technology really to blame? Humans are always looking for a scapegoat. We want someone, or something, to blame for our problems when really the problems we have in our world are always human problems. It's always the same old wolves in brand new sheepskin. Pain, suffering, loss, dishonesty, theft, divorce, unfaithfulness, loneliness, abuse, laziness, etc. It's the same old circle of bad guys that cause all our problems. In a word, it's sin that causes our issues.

Here's something that is absolutely important for evangelists to understand. Sin is usually believed to be simply the destruction of good and it is that, sort of. A more accurate representation of sin is the binary representation of one and zero. Good is the existence of God, of love, and sin is the absence of good. In Spanish, sin means "without". In Hebrew, the word for sin means to miss the mark. It's an archer's term that basically means that you missed the target you were shooting for. Those are important definitions as they describe that something is missing and the person who sinned was actually shooting for the correct target and missed. Much like I described in the last chapter, you can recognize that someone was aiming for something good even if they missed their target. The issue with technology is that people are missing the target that they are after. Some miss by a long shot while others get extremely close. If we spend all our time as evangelists demonizing technology, we will lose out on the opportunity to invite others into the beauty of our faith and they will stay insistent that there is a great truth to uncover in technology. They're right. Rather than push them to indulgence by harping on repression, we should be looking to enter into their world as well so that they can help illuminate our faith through their discoveries in technology. As an evangelist it's important to hold fast to the truth that you do not know everything. Someone who has never stepped foot in a church can teach you something about truth and you must always be open to listening to them. Saint John Paul II was a master at recognizing where people saw the truth of the faith in the world. He worked hard to demonstrate to them that their desires for good, were actually good! The trick is to help people understand that while desires for good things are good, the way we go about getting the good that we desire is not always good. An inspirational leader seeks to show the way to the good that people seek, not to simply chastise people for the times that they miss the mark. I am a strong personality and I am not afraid to tell people things that are uncomfortable. Still, I

recognize that this method must be used carefully
and strategically and must be coupled with mercy,
love, and inspirational guidance. Don't simply tell
people that their attachment to the symbol is
unhealthy. Show them what the symbol is meant to
signify and then give them the freedom to pursue it
or to choose not to. As evangelists, our job is to be a
sign ourselves. We must live out the principles that
we preach in order to ensure that our words have
the value that they should. Like God, we must
present the invitation and respect the freedom of
the person to make their own choice. It is their
journey, not yours.

To those who picked up this book because you were
curious how technology could reveal God, I hope I
gave you some ideas to chew on. I am one person
and I don't have all the answers. Computer
programmers do not build everything from scratch.
My knowledge in computers and theology that
helped me write this book comes from many
different sources over the years. I've read many
books, been in countless conversations on these
topics, read articles, watched videos, went to
conferences, took classes and worked in jobs where
I learned many of the things that I know. In
computer science, a programmer has access to code
libraries where other programmers before them
create libraries of functions that a programmer can
pull from to use and then build on top of them. As
people create more complex functions that may be
of use to others, they get added to code libraries, or
central repositories like GitHub, to be shared with

others. I'm big on open source technologies and sharing. Let me tell you, there is a vast amount of knowledge in the "code libraries" of the Catholic Church. Along with the Bible, the traditions of the Catholic Church (which I recognize can cause Protestants to recoil if they don't understand them) are like code libraries where people have developed prayers, writings, and practices, that serve to illuminate the faith of others.

If you are hearing all of this stuff for the first time let me recommend some books to you that you might enjoy. Start with Frank Sheed's *Theology and Sanity* or Christopher West's *Theology of the Body for Beginners*. In no particular order, I recommend *The Last Superstition* by Edward Feser, *The Godless Delusion* by Patrick Madrid and Kenneth Hensley, *Handbook of Christian Apologetics* by Peter Kreeft, *Jesus of Nazareth* by Pope Benedict XVI. I recommend books or videos from Archbishop Fulton Sheen, Father Robert Spitzer, Scott Hahn and absolutely anything written by Pope John Paul II (to include his *Theology of the Body* and *Love and Responsibility*). There are so many great authors and apologists, these are only some of the modern ones. St. Thomas Aquinas and St. Augustine are some amazing old school intellectuals and of course the Bible is the main book that should be at the source of any serious study. If you want to get a great introduction into better understanding the Bible, I highly recommend Jeff Cavin's *Bible Timeline*. It will illuminate the story of salvation for you and help you to understand how to read it. I never set out to completely cover every area of this topic. My goal was to start the fire and offer my insights to the community. I would love it if people smarter than me took this information and ran with it. Please, draw out these concepts and leverage your knowledge and experience to help me build this bridge. My desire is that techies and non-techies will take the opportunity to extend themselves outside of their comfort zone and come to learn about each other. I believe deep in my heart that people who have a deep love for technology are searching for the creator of technology. Logic himself. More than that, I believe that people who have a deep understanding of technology have a lot to teach people like me. I am thirsting for truth. I want to learn, and I know that there are people out there like Peter Jaros who have stumbled upon great truths in their passion for learning about technology. I want to hear what they have to say. Like Johnny Five, I need more input! I desire to continue hacking theology by learning more about technology and the way that it represents the great truth of the world.

Computer programmers often learn to code their first program by writing code to display the text, "Hello World!". It's become a popular first program because it's simple to develop but it's also meaningful because the programmer is entering a new world of possibilities. My hope is that, for you, this is just the beginning; an introduction to a new world. This is your "Hello, World!" program. Where you go from here is up to you. The possibilities are endless. Be a hacker. Don't settle. Your desire for technology means something. Look deeper, keep pushing. Break things. Make mistakes. Open your mind and keep searching. Leverage the community, don't try and do it all yourself. Share your findings. I mentioned at the beginning that I was a kid who loved to stay up all hours of the night on the computer and often I had to hide from my parents so that I could get extra time.

I now have a six-year-old son who is as in love with technology as I am. He fascinates me with his passion for technology. When he was only three and a half years old, he would often try to get on the computer and I wouldn't let him. He was so drawn to the computer that he somehow started waking himself up at 5 AM, before my wife and I would wake up, to get on the computer before we could stop him. At the time I never locked it because I never had reason to. The first day I caught him, I got mad at him and turned off the screen before I went to bed. The next morning, I woke up and he was on the computer. He had turned the screen on. Not a difficult task but not something I expected from a three-year-old. I got mad at him again and that night I unplugged the power cable from the back of the screen. The next morning, I woke up and he was standing on the chair, looking at the back of the computer monitor trying to figure out why it wasn't turning on. I almost got frustrated that he was so persistent until I realized that what he was doing was actually amazing. I recognized that he had the mind of a hacker. Every time I tried to make the problem more difficult for him, he figured out a way to solve the problem and get beyond my preventative measures. I recognized that day that I needed to protect him from the dangers of technology, but I also needed to affirm his love for it even more. That was three years ago. The realization that even a three-year-old could recognize that the computer contained the promise of some great truth helped set me on the path to eventually write this book. I was once that little kid and I see a lot of myself in him. I hope that, for all the kids like him, and like I once was, we never lose hope that the next generation will uncover those truths and take them to greater heights. I have great expectations for my son and for all you lovers of technology. I can't wait to see what the future brings.

Hello, World!

REFERENCES

[1] Daniel R. Headrick, Technology: A World History (New York, NY: Oxford University Press, Inc., 2009), 6, 16, 20.

[2] Etymonline.com. Technology. Retrieved from https://www.etymonline.com/word/technology.

[3] Whipps, Heather. (2008, March 02). How the Iron Age Changed the World. Retrieved from https://www.livescience.com.

[4] Michael R. Swaine and Paul A. Freiberger. (2017, June 28). Difference Engine. https://www.britannica.com/technology/Difference-Engine.

[5] Census.gov. The Hollerith Machine. Retrieved from https://www.census.gov/history/www/innovations/technology/the_hollerith_tabulator.html

[6] McMenemy, Rachael. (2017, November 01). Stephen Hawking says he fears artificial intelligence will replace humans. Retrieved from https://www.cambridge-news.co.uk.

[7] Christ In Prophecy. "Nathan Jones on the Sign of Technology." YouTube. YouTube, 08 November 2013. Web. 01 September 2017.

[8] Miller, Ron. (2015, March 21). Anti-Tech Backlash Could Be Coming Soon to A City Near You. Retrieved from https://techcrunch.com/.

[9] History.com Staff. (2009). Industrial Revolution. Retrieved from http://www.history.com/.

[10] The Social Impact of the Industrial Revolution." Science and Its Times: Understanding the Social Significance of Scientific Discovery. Encyclopedia.com. 5 Mar. 2018 <http://www.encyclopedia.com>.

[11] Baraniuk, Chris. (2018). Exclusive: UK police wants AI to stop violent crime before it happens. Retrieved from https://www.newscientist.com/article/2186512-exclusive-uk-police-wants-ai-to-stop-violent-crime-before-it-happens/.

[12] Rosenwald, Michael. (2017). How Hugh Hefner became Hef: From sexually repressed upbringing to renowned Playboy. Retrieved from https://www.washingtonpost.com/news/retropolis/wp/2017/09/28/how-hugh-hefner-became-hef-from-sexually-repressed-childhood-to-playboy.

[13] Peeja. (2014, June 23). Why do computers use binary, anyway? — The Binary Tree [Video file]. Retrieved from https://www.youtube.com/watch?v=1sWCBgGALXE.